Understanding AC Circuits

Understanding AC Circuits

Dale R. Patrick

Stephen W. Fardo

Newnes

Boston Oxford Auckland Johannesburg Melbourne New Delhi

Newnes is an imprint of Butterworth–Heinemann.

Copyright © 2000 by Butterworth–Heinemann

 A member of the Reed Elsevier group

All rights reserved.

No part of this publication may be reproduced, stored in a retrieval system, or transmitted in any form or by any means, electronic, mechanical, photocopying, recording, or otherwise, without the prior written permission of the publisher.

♾ Recognizing the importance of preserving what has been written, Butterworth–Heinemann prints its books on acid-free paper whenever possible.

GLOBAL RELEAF Butterworth–Heinemann supports the efforts of American Forests and the Global ReLeaf program in its campaign for the betterment of trees, forests, and our environment.

Library of Congress Cataloging-in-Publication Data

Patrick, Dale R.
　　Understanding AC Circuits / by Dale R. Patrick and Stephen W. Fardo.
　　　　　　p. cm.
　　ISBN 0-7506-7103-3 (alk. paper)
　　1. Electric circuits—Alternating current.　I. Fardo, Stephen W.
II. Title.
TK7816.P35 1999
621.319′13—dc21　　　　　　　　　　　　　　　　99-16982
　　　　　　　　　　　　　　　　　　　　　　　　　CIP

British Library Cataloguing-in-Publication Data

A catalogue record for this book is available from the British Library.

The publisher offers special discounts on bulk orders of this book.
For information, please contact:
Manager of Special Sales
Butterworth-Heinemann
225 Wildwood Avenue
Woburn, MA 01801-2041
Tel: 781-904-2500
Fax: 781-904-2620

For information on all Newnes publications available, contact our World Wide Web home page at: http://www.newnespress.com

10 9 8 7 6 5 4 3 2 1

Printed in the United States of America

Contents

PREFACE		ix
COURSE OBJECTIVES		xi
PARTS LIST FOR EXPERIMENTS		xiii
UNIT ONE	**BASICS OF ALTERNATING CURRENT (AC)**	**1**
	Unit Objectives	1
	Important Terms	2
	Alternating Current (AC) Voltage	3
	Single-Phase and Three-Phase AC	4
	Electromagnetic Induction	5
	Generating AC Voltage	6
	Basics of Electrical Generators	7
	Single-Phase AC Generation	7
	Three-Phase AC Generation	8
	Self-Examination	10
	Unit 1 Examination: Basics of Alternating Current (AC)	12
UNIT TWO	**MEASURING AC**	**15**
	Unit Objectives	15
	Important Terms	16
	Measuring AC Voltage with a Multimeter	17
	Measuring AC Voltage with an Oscilloscope	17
	Oscilloscope Operation	18
	Self-Examination	33

Experimental Activities for AC Circuits	34
Tools and Equipment	35
Important Information	35
Lab Activity Troubleshooting and Testing	36
Experiment 2-1—Measuring AC Voltage	40
Experiment 2-2—Measuring AC with an Oscilloscope	44
Unit 2 Examination: Measuring AC	48

UNIT THREE — RESISTANCE, INDUCTANCE, AND CAPACITANCE IN AC CIRCUITS — 51

Unit Objectives	51
Important Terms	52
Resistive AC Circuits	54
Inductive Circuits	55
Capacitive Circuits	58
Vector (Phasor) Diagrams	60
Series AC Circuits	61
Parallel AC Circuits	63
Power in AC Circuits	66
Self-Examination	72
Experiment 3-1—Inductance and Inductive Reactance	78
Experiment 3-2—Capacitance and Capacitive Reactance	81
Experiment 3-3—Series RL Circuits	84
Experiment 3-4—Series RC Circuits	87
Experiment 3-5—Series RLC Circuits	89
Experiment 3-6—Parallel RL Circuits	92
Experiment 3-7—Parallel RC Circuits	95
Unit 3 Examination: Inductance and Capacitance in AC Circuits	98

UNIT FOUR — TRANSFORMERS — 103

Unit Objectives	103
Important Terms	104
Transformer Operation	105
Types of Transformers	105
Transformer Efficiency	107
Self-Examination	107
Experiment 4-1—Transformer Analysis	110
Unit 4 Examination: Transformers	112

UNIT FIVE	**FREQUENCY-SENSITIVE AC CIRCUITS**	**115**
	Unit Objectives	115
	Important Terms	116
	Filter Circuits	117
	Resonant Circuits	118
	Self-Examination	123
	Decibels	124
	Logarithms	124
	Decibel Applications	125
	Using Decibels with Filter Circuits	127
	Waveshaping Control	134
	Differentiator Circuits	135
	Integrator Circuits	136
	Self-Examination	136
	Experiment 5-1—Low-Pass Filter Circuits	139
	Experiment 5-2—High-Pass Filter Circuits	142
	Experiment 5-3—Band-Pass Filter Circuit	145
	Experiment 5-4—Series Resonant Circuits	147
	Experiment 5-5—Parallel Resonant Circuits	150
	Unit Five Examination: Frequency-Sensitive AC Circuits	152
APPENDIX A	**ELECTRONICS SYMBOLS**	**155**
APPENDIX B	**TRIGONOMETRY FOR AC ELECTRONICS**	**159**
APPENDIX C	**ELECTRONIC EQUIPMENT AND PARTS SALES**	**161**
APPENDIX D	**SOLDERING TECHNIQUES**	**163**
APPENDIX E	**TROUBLESHOOTING**	**165**
APPENDIX F	**USE OF A CALCULATOR**	**167**
	INDEX	**169**

Preface

Understanding AC Circuits is an introductory text that provides coverage of the various topics in the field of alternating current (ac) electronics. The key concepts in this book are discussed through a simplified approach that greatly enhances learning. The use of mathematics is discussed through applications and illustrations.

Every unit is organized in a step-by-step progression of concepts and theory. Each unit begins with an *introduction* and *unit objectives*. A discussion of important concepts and theories follows. A *self-examination*, with answers provided, is integrated into each chapter to reinforce learning. *Experimental activities*, with components and equipment listed, are included in each unit to help students learn electronics through practical experimental applications. The final learning activity for each unit is a *unit examination*, which includes at least twenty objective, multiple-choice questions.

Definitions of *important terms* are presented at the beginning of each unit. Several *appendices* are used to aid students in performing experimental activities. The expense of the equipment needed for the experiments is kept to a minimum. A comprehensive parts list is provided, as is information on electronics distributors. The experiments suggested are low-cost activities that can be performed in the home or a school laboratory. They are very easy to understand and emphasize troubleshooting concepts. Electronics can be learned experimentally at low cost by means of completion of these lab activities. Appendices dealing with electronics symbols and soldering are provided for easy reference.

This book covers alternating current (ac) circuits, an important foundation in the study of electronics. The companion text is *Understanding DC Circuits*. Both of these books are organized in the same easy-to-understand format. They can be used to acquire a basic understanding of electronics in the home, school, or workplace. The sequence of the books allows the student to progress at a desired pace in the study of electronics basics and to perform experiments with low-cost equipment and supplies (depending on availability). As students progress, they may wish to purchase various types of test equipment at varying degrees of expense. The experiments allow the students to further develop an understanding of the topics discussed in each unit. They are intended as an important supplement to student learning.

The following supplemental materials are available to provide an aid to effective learning:

1. *Instructor's Resource Manual*—answers to all unit examinations and suggested data for experimental activities, including a comprehensive analysis of each experiment.

2. *Instructor's Transparency Masters*—enlarged reproductions of selected illustrations in the textbook that are suitable for use for transparency preparation for class presentations.

3. *Instructor's Test Item File*—suggested objective, multiple-choice questions for use with each unit of instruction.

These supplements are extremely valuable for instructors in organizing electronics classes. The complete instructional cycle, from objectives to evaluation, is included in this series of books. We hope you will find this book easy to understand and that you are successful in your pursuit of knowledge in an exciting technical area. Electronics is an extremely vast and interesting field of study. *Understanding AC Circuits* provides a foundation for understanding electronics technology.

Dale R. Patrick
Stephen W. Fardo
Richmond, Kentucky

Course Objectives

Upon completion of this course on understanding ac circuits, you should be able to do the following:

1. Understand the following basic ac electronics concepts:
   a. Single-phase ac
   b. Three-phase ac
   c. Electromagnetic induction
   d. AC sine wave voltage.

2. Use a VOM or oscilloscope to measure ac quantities.

3. Solve basic electronics problems with ac circuits that involve capacitance, inductance, impedance, reactance, resistance, and power calculations.

4. Explain the operation of transformers in ac circuits.

5. Describe the operation of frequency-sensitive filter circuits and resonant circuits.

6. Construct experimental ac circuits using schematics and perform tests and measurements with a multimeter and signal generator.

Parts List for Experiments

Various components and equipment are needed to perform the experimental activities in this course. These parts may be obtained from electronics suppliers, mail-order warehouses, or educational supply vendors. A list of several of these is included in Appendix C. The following equipment and components are necessary for the successful completion of the activities included in this course.

RESISTORS (1 of each)

100 Ω, ½ W, 5%

220 Ω, ½ W, 5%

300 Ω, ½ W, 5%

1000 Ω, ½ W, 5%

1200 Ω, ½ W, 5%

10 kΩ, ½ W, 5%

22 kΩ, ½ W, 5%

CAPACITORS (1 of each)

0.01 µF, 35 V

10 µF, 35 V

0.05 µF, 35 V

0.005 µF, 35 V

MISCELLANEOUS

Solder package

10 feet of no. 22 solid wire

6 V light bulb

Light socket

Slide switch (DPDT)

4.5-henry inductor

Connecting wires

Multimeter—analog or digital

Oscilloscope

Audio signal generator

6 V battery

30 V ac source

Small parts container

UNIT 1

Basics of Alternating Current

Alternating current (ac) electronics is somewhat more complex than direct current (dc) electronics. AC circuits, like dc circuits, have a *source* of energy and a *load* in which power conversion takes place. Most of the electric energy produced in the United States is alternating current; therefore, ac systems are used for many applications. In terms of electronic circuits, three important characteristics are present in ac circuits. These characteristics are *resistance, inductance,* and *capacitance.* There are two types of ac *voltage*: single-phase and three-phase. These are discussed in the units that follow.

UNIT OBJECTIVES

Upon completion of this unit you will be able to do the following:

1. Explain the difference between ac and dc.
2. Define the process of electromagnetic induction.
3. Describe factors that affect induced voltage.
4. Draw a simple ac generator and explain ac voltage generation.
5. Convert peak, peak-peak, average, and RMS-effective values from one to the other.
6. Explain the relation between period and frequency of an ac waveform.
7. Recognize the different types of ac waveforms.

Important Terms

The following terms provide a review of the basics of ac electronics:

Average voltage (V_{avg}) The value of an ac sine wave voltage; found with the formula $V_{avg} = V_{peak} \times 0.636$.

Cycle A sequence of events that causes one complete *sine wave* or pattern of alternating current. It begins from a zero reference and goes in a positive direction, back to zero, in a negative direction, and back to zero, a complete 360° sequence.

Delta connection A method of connecting three-phase circuits in which the beginning of one phase is connected to the end of the adjacent phase.

Effective voltage (V_{eff}) A value of an ac sine wave voltage that has the same effect as an equal value of dc voltage; also called RMS (root mean square) voltage; $V_{eff} = V_{peak} \times 0.707$.

Frequency (f) The number of ac cycles per second; measured in hertz (Hz).

Hertz The international unit of measurement of frequency; equal to one cycle per second (cps).

In-phase The condition in which two ac waveforms of the same frequency pass through their minimum values at the same time and same polarity.

Instantaneous voltage (V_i) A value of ac voltage at a given instant of time along a waveform.

Peak voltage (V_{peak}) The maximum positive or negative value of an ac sine wave voltage; $V_{peak} = V_{eff} \times 1.41$.

Peak-to-peak voltage ($V_{p\text{-}p}$) The value of an ac sine wave voltage from its positive peak to its negative peak.

Period (time) The time required to complete one ac cycle; time = 1/frequency.

Phase angle (θ) The angular displacement (in degrees) between applied voltage and total current flow in an ac circuit.

Root mean square (RMS) voltage Same as effective voltage.

Sine wave A waveform that represents one cycle of ac voltage; see *cycle*.

Single-phase ac The voltage output produced by a single-phase generator in the form of a series of sine waves.

Theta (θ) The Greek letter used to represent the phase angle of an ac circuit.

Three-phase ac The voltage produced by a three-phase generator in the form of a series of three sine waves separated in phase by an angle of 120°.

Waveform The pattern of an ac frequency derived by looking at instantaneous voltage values that occur over a period of time. A waveform is plotted on a graph with instantaneous voltages on the vertical axis and time on the horizontal axis.

Wavelength The distance from a point on a waveform to a corresponding point on an adjacent waveform.

Wye connection A method of connecting three-phase circuits in which the beginnings or ends of each phase are connected together to form a common or neutral point.

Alternating Current (AC) Voltage

When an ac source is connected to some type of load, the direction of the current changes several times in a given unit of time. Remember that direct current (dc) flows in one direction only. A diagram of one cycle of ac is compared with a dc waveform in Fig. 1-1. This waveform is called an *ac sine wave*. When an ac generator shaft rotates one complete revolution, or 360°, one ac sine wave is produced. AC voltage generation is discussed at the end of this unit. The sine wave has a positive peak at 90° and decreases to zero at 180°. It increases to a peak negative voltage at 270° and decreases to zero at 360°. The cycle repeats itself. Current flows in one direction during the positive alternation and in the opposite direction during the negative half-cycle.

Figure 1-2 shows five cycles of ac. If the time required for an ac generator to produce five cycles were 1 second, the frequency of the ac would be 5 cycles per second (cps). AC generators at power plants in the United States operate at a frequency of 60 cps, or 60 hertz (Hz). The hertz is the international unit for frequency measurement. If 60 ac sine waves are produced every second, a speed of 60 rev/s is needed. This produces a frequency of 60 cps.

AC voltage is measured with a volt-ohm-milliammeter (VOM), also called a *multimeter*. Polarity of the meter leads is not important. This is because ac changes direction. Polarity is important, however, in measuring dc, because the current flows only in one direction. Some VOMs cannot be used to measure ac current. They have ranges for ac voltage only.

Figure 1-3 shows several voltage values associated with ac. Among these are peak positive, peak negative, and peak-to-peak ac values. *Peak positive* is the maximum positive voltage reached during a cycle of ac. *Peak negative* is the maximum negative voltage reached. *Peak-to-peak* is the voltage value from peak positive to peak negative. These values are important to know for work with radio and television amplifier circuits. For example, the most important ac value is called *effective*, or *measured*, value. This value is less than the peak positive value. A common ac voltage is 120 V, which is used in homes. This is an effective value voltage. Its peak value is approximately 170 V.

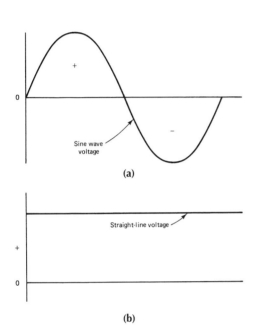

Fig. 1-1. Comparison of ac (a) and dc (b) waveforms.

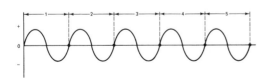

Fig. 1-2. Five cycles of alternating current.

The effective value of ac is defined as the ac voltage that will do the same amount of work as a dc voltage of the same value. For example, in the circuit in Fig. 1-4 if the switch is placed in position 2, a 10 V ac effective value is applied to the lamp. The lamp should produce the same amount of brightness with a 10 V ac effective value as with 10 V dc applied. When ac voltage is measured with a meter, the reading is effective value.

In some cases it is important to convert one ac value to another. For instance, the voltage rating of electronic devices must be greater than the peak ac voltage applied to them. If 120 V ac is the measured voltage applied to a device, the peak voltage is about 170 V. The device must be rated over 170 V rather than 120 V.

To determine peak ac when the measured or effective value is known, the following formula is used:

$$\text{Peak} = 1.41 \times \text{effective value}$$

When 120 V is multiplied by the 1.41 conversion factor, the peak voltage is found to be about 170 V.

Two other terms that should be mentioned are RMS value and average value. RMS stands for *root mean square*, and it is equal to 0.707 × peak value. RMS refers to the mathematical method used to determine effective voltage. RMS voltage and effective voltage are the same. Average voltage is the mathematical average of all instantaneous voltages that occur at each period of time throughout an alternation. The average value is equal to 0.636 times the peak value.

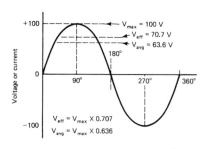

Fig. 1-3. Voltage values of an ac waveform.

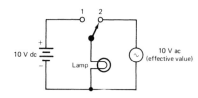

Fig. 1-4. Comparison of effective ac voltage and dc voltage.

Single-Phase and Three-Phase AC

Single-phase ac voltage is produced by single-phase ac generators, or it can be obtained across two power lines of a three-phase system. A single-phase ac source has a hot wire and a neutral wire. The neutral wire is grounded to help prevent electric shocks. Single-phase power is the type of power distributed to homes. A three-phase ac source has three power lines. Three-phase voltage is produced by three-phase generators at power plants. Three-phase voltage is a combination of three single-phase voltages connected together electrically. This voltage is similar to three single-phase sine waves separated in phase by 120°. Three-phase ac is used to power large equipment in industry and commercial buildings. It is not distributed to homes. There are three power lines on a three-phase system. Some three-phase systems have a neutral connection, and others do not.

The term *phase* refers to time or the difference between one point and another. If two sine-wave voltages reach their zero and maximum values at the same time, they are in phase. Figure 1-5 shows two alternating current voltages that are in phase. If two voltages reach their zero and maximum values at different times, they are out of phase. Figure 1-6 shows two ac voltages that are

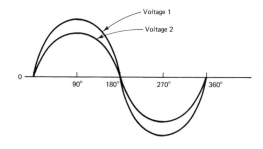

Fig. 1-5. Two ac voltages that are in phase.

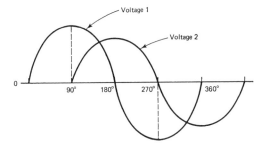

Fig. 1-6. Two ac voltages that are out of phase at an angle of 90°.

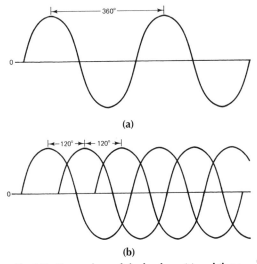

Fig. 1-7. Comparison of single-phase (a) and three-phase (b) voltages.

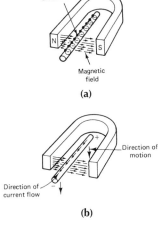

Fig. 1-8. Principle of electromagnetic induction. (a) Conductor placed inside a magnetic field. (b) Conductor moved downward through the magnetic field.

out of phase. Phase difference is given in degrees. The voltages shown are out of phase at an angle of 90°.

Single-phase ac voltage is in the form of a sine wave. Single-phase ac voltage is used for low-power applications, primarily in homes. Almost all electric power is generated and transmitted over long distances as three-phase ac. Three coils are placed 120° apart in a generator to produce three-phase ac voltage. Most ac motors larger than 1 horsepower (hp) in size operate with three-phase ac power applied.

Three-phase ac systems have several advantages over single-phase systems. In a single-phase system, the power is said to be pulsating. The peak values along a single-phase ac sine wave are separated by 360°, as shown in Fig. 1-7a. This is similar to a one-cylinder gas engine. A three-phase system is somewhat like a multicylinder gas engine. The power is steadier. One cylinder is compressing when the others are not. This is similar to the voltages in three-phase ac systems. The power of one separate phase is pulsating, but the total power is constant. The peak values of three-phase ac are separated by 120° (Fig. 1-7b). This makes three-phase ac power more desirable to use.

The power ratings of motors and generators are greater when three-phase ac power is used. For a certain frame size, the rating of a three-phase ac motor is almost 50% larger than a similar single-phase ac motor.

Electromagnetic Induction

AC electric energy is produced by means of placing a conductor inside a magnetic field. An experiment by a scientist named Michael Faraday was very important. It showed the following principle: When a conductor moves across the lines of force of a magnetic field, electrons in the conductor tend to flow in a certain direction. When the conductor moves across the lines of force in the opposite direction, electrons in the conductor tend to flow in the opposite direction. This is the principle of generation of electric power. Most of the electric energy used today is produced with magnetic energy. Figure 1-8 shows the principle of electromagnetic induction. Electric current is produced only when there is motion. When the conductor is brought to a stop while crossing lines of force, electric current stops.

If a conductor or a group of conductors is moved through a strong magnetic field, induced current flows and a voltage are produced. Figure 1-9 shows a loop of wire rotated through a magnetic field. The position of the loop inside the magnetic field determines the amount of induced current and voltage. The opposite sides of the loop move across the magnetic lines of force in opposite directions. This movement causes an equal amount of electric current to flow in *opposite directions* through the two sides of the loop. Notice each position of the loop and the resulting output voltage in Fig. 1-9. The electric current flows in one direction and then in the opposite direction with every complete revolution of the conductor. This method produces ac. One complete rotation is called a cycle. The number of cycles per second is known as the *frequency*. Most ac generators produce 60 cps.

The ends of the conductor that move across the magnetic field of the generator in Fig. 1-9 are connected to slip rings and brushes. The slip rings are mounted on the same shaft as the conductor. Carbon brushes are used to make contact with the slip rings. The electric current induced into the conductor flows through the slip rings to the brushes. When the conductor turns half a revolution, electric current flows in one direction through the slip rings and the meter. During the next half revolution of the coil, the positions of the two sides of the conductor are opposite. The direction of the induced current is reversed. Current now flows through the meter in the opposite direction.

The conductors that make up the rotor of a generator have many turns. The voltage generated is determined by the number of turns of wire used, the strength of the magnetic field, and the speed of the prime mover used to rotate the machine.

Direct current also can be produced through electromagnetic induction. A simple dc generator has a *split-ring commutator* rather than two slip rings. The split rings resemble one full ring except that they are separated by small openings. Induced electric current still flows in opposite directions to each half of the split ring. However, current flows in the same direction in the load circuit because of the action of the split rings.

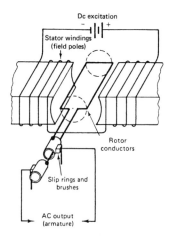

Fig. 1-9. Basic ac generator.

Generating AC Voltage

Electromagnetic induction takes place when a conductor passes through a magnetic field and cuts across lines of force. As a conductor is passed through a magnetic field, it cuts across the magnetic flux lines. As the conductor cuts across the flux lines, the magnetic field develops a force on the electrons of the conductor. The direction of the electron movement determines the polarity of the induced voltage. The *left-hand rule* is used to determine the direction of electron flow. This rule for generators is stated as follows: Hold the thumb, forefinger, and middle finger of the left hand perpendicular to each other. Point the forefinger in the direction of the magnetic field from north to south. Point the thumb in the direction of the motion of the conductor. The middle finger will then point in the direction of electron current flow.

The amount of voltage induced into a conductor cutting across a magnetic field depends on the number of lines of force cut in a given time. This is determined by the following three factors:

1. The *speed* of the relative motion between the magnetic field and the conductors
2. The *strength* of the magnetic field

3. The *length* of the conductor passed through the magnetic field

If the speed of the conductor cutting the magnetic lines of force is increased, the generated voltage increases. If the strength of the magnetic field is increased, the induced voltage also is increased. A longer conductor allows the magnetic field to induce more voltage into the conductor. The induced voltage increases when each of the three quantities is increased.

Basics of Electric Generators

Electric generators are used to produce electric energy. They require some form of mechanical energy input. The mechanical energy is used to move electric conductors (turns of wire) through a magnetic field inside the generator. All generators operate by means of electromagnetic induction. A generator has a stationary part and a rotating part housed inside a machine assembly. The stationary part is called the *stator* and the rotating part is called the *rotor*. The generator has north and south *magnetic field poles*. Generators must have a method of producing rotary motion (mechanical energy). This system is called a *prime mover* and is connected to the generator shaft. There must also be a method of electrically connecting the rotating conductors to an external circuit. This is done with a slip ring or split ring and brush assembly. *Slip rings* are used with ac generators, and *split rings* are used with dc generators. Slip rings are made of solid sections of copper; split rings are made of several copper sections that are separated from each other. The rings are permanently mounted on the shaft of a generator and connected to the ends of the conductors of the rotor. When a load is connected to a generator, a complete circuit is made. With all generator parts working together, electric power is produced. A basic ac generator is shown in Fig. 1-9.

Single-Phase AC Generation

Single-phase electric power is often used, particularly in homes. However, single-phase generators produce very little electric power. AC generators usually are called *alternators*. Single-phase electric power used in homes usually is produced by three-phase generators at power plants. It is converted to single-phase electric energy before it is distributed to homes.

The current produced by single-phase generators is in the form of a *sine wave*. This waveform is called a sine wave because of its mathematical origin. It is based on the trigonometric sine function used in mathematics (see appendix B).

The voltage induced into the conductors of an armature varies as the sine of the angle of rotation between the conductors and the magnetic field (Fig. 1-10). The voltage induced at a specific time is called *instantaneous voltage* (V_i). Voltage induced into an armature conductor at a specific time is found by using the following formula:

$$V_i = V_{max} \times \sin \theta$$

V_{max} is the maximum voltage induced into the conductor. The symbol theta (θ) is the angle of conductor rotation. For example, at the 60° position, assume that the maximum voltage output is 100 V. The instantaneous voltage induced at 60° is $V_i = 100 \times \sin \theta$. It may be necessary to study appendix B at this time. The sine of 60° is 0.866, so induced voltage at 60° is 86.6 V ($V_i = 100 \times 0.866 = 86.6$ V). A calculator with trigonometric functions can be used (see appendix F).

The frequency of the voltage produced by alternators is usually 60 Hz. A *cycle* of ac is generated when the rotor moves one complete revolution (360°). *Cycles per second* or *hertz* refers to the number of revolutions per second. For example, a speed of 60 rev/s (3600 rev/min) produces a frequency of 60 Hz. The frequency (f) of an alternator is found with the following formula:

$$f(Hz) = \frac{\text{number of magnetic poles} \times \text{speed of rotation (rev/min)}}{120}$$

The frequency is measured in hertz. If the number of poles (field coils) is increased, the speed of rotation can be reduced and still produce a 60-Hz frequency.

For a generator to convert mechanical energy into electric energy, the following three conditions must exist: (1) there must be a magnetic field, (2) there must be conductors cutting the magnetic field, and (3) there must be relative motion between the magnetic field and the conductors.

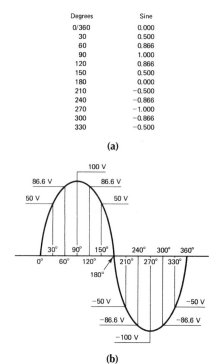

Fig. 1-10. Generation of an ac sine wave: (a) sine values of angles from 0 to 360°; (b) sine wave produced.

Three-Phase AC Generation

Most electric power produced in the United States is three-phase ac produced at power plants. Power-distribution systems use many three-phase generators (alternators) connected in parallel. A simple diagram of a three-phase alternator and a three-phase voltage diagram are shown in Fig. 1-11. The alternator output windings are connected in either of two ways. These methods are called the *wye connection* and the *delta connection*. These three-phase connections are shown with schematics in Fig. 1-12. These methods also are used for connecting the windings of three-phase transformers, three-phase motors, and other three-phase equipment.

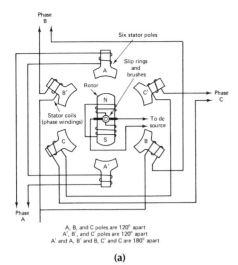

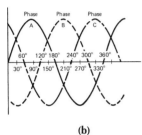

Fig. 1-11. (a) Diagram of a three-phase alternator; (b) diagram of three-phase voltage.

The differences between voltages and currents in wye and delta systems should be remembered. In the *wye connection* of Fig. 1-12a, the beginnings or ends of each winding are connected together. The other sides of the windings are the ac lines that extend from the alternator. The voltage across the power lines is called *line voltage* (V_L). Line voltage is higher than the voltage across each phase. Line voltage equals $\sqrt{3}$ (1.73) multiplied by the voltage across the phase windings (V_p), so $V_L = V_p \times 1.73$. The line current (I_L) equals the phase current (I_p), or $I_L = I_p$. In the *delta connection* (see Fig. 1-12b), the end of one phase winding is connected to the beginning of the next phase winding. Line voltage (V_L) is equal to the phase current (I_p) multiplied by 1.73, so $I_L = I_p \times 1.73$.

Three-phase power is used mainly for high-power industrial and commercial equipment. The power produced by three-phase generators is a more constant output than single-phase power. Three separate single-phase voltages can be delivered from a three-phase power system. It is more economical to distribute three-phase power from power plants to homes, cities, and industries. Three conductors are needed to distribute three-phase voltage, whereas six conductors are necessary for three separate single-phase systems. Equipment that uses three-phase power is smaller than similar single-phase equipment. It saves energy to use three-phase power whenever possible.

One type of a three-phase alternator is used in automobiles. The three-phase ac it produces is converted to dc by a rectifier circuit. The dc voltage is used to charge the automobile battery. Charging time and voltage are controlled by a voltage-regulator circuit.

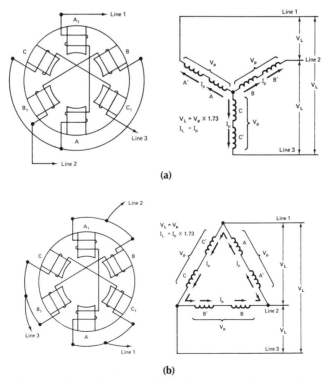

Fig. 1-12. Three-phase connections: (a) wye connection, sometimes called star connection; (b) delta connection.

Basics of Alternating Current 9

Self-Examination

Several basic ac values that are commonly used in electric measurement and problem solving. These units are derived from ac sine-wave relationships. The following ac sine-wave values are often covered from one value to the other:

Effective value (RMS) = 0.707 × peak value

Average value = 0.636 × peak value

Peak value = 1.41 × RMS value

Peak-to-peak value = 2.82 × RMS value

To complete this self-test, you should review ac sine-wave relationships.

Instructions: Solve the following problems by placing the correct answer in the blank spaces.

Compute the following effective values. Values given are peak voltage:

Peak

1. 4 V ac = _____ RMS
2. 12 V ac = _____ RMS
3. 6 V ac = _____ RMS
4. 18 V ac = _____ RMS
5. 15 V ac = _____ RMS
6. 2 V ac = _____ RMS
7. 5 V ac = _____ RMS
8. 9 V ac = _____ RMS

Compute the following peak and peak-to-peak (p-p) values. Values given are RMS voltage.

RMS

9. 3 V ac = _____ peak; _____ p-p
10. 8 V ac = _____ peak; _____ p-p
11. 7 V ac = _____ peak; _____ p-p
12. 9 V ac = _____ peak; _____ p-p
13. 10 V ac = _____ peak; _____ p-p
14. 15 V ac = _____ peak; _____ p-p
15. 11 V ac = _____ peak; _____ p-p
16. 18 V ac = _____ peak; _____ p-p

Complete the following.

17. Alternating current is defined as electron flow that is constantly changing in _____.

18. One complete alternation of alternating voltage or current is called a _____.

19. A cycle of alternating voltage or current is represented by a _____.
20. One complete cycle is _____ degrees.
21. An alternation is _____ degrees.
22. The part of a sine wave above the horizontal time line is called the _____ alternation.
23. The part of a sine wave below the horizontal time line is called the _____ alternation.
24. The number of cycles per second is called _____.

Answers

1.	2.82 RMS	2.	8.48 RMS
3.	4.24 RMS	4.	7.07 RMS
5.	1.06 RMS	6.	1.414 RMS
7.	3.53 RMS	8.	6.363 RMS
9.	4.23 peak; 8.46 p-p	10.	11.28 peak; 22.56 p-p
11.	9.87 peak; 19.74 p-p	12.	12.69 peak; 25.38 p-p
13.	14.1 peak; 28.2 p-p	14.	21.15 peak; 42.3 p-p
15.	15.51 peak; 31.02 p-p	16.	25.38 peak; 50.76 p-p
17.	Direction or magnitude	18.	Cycle
19.	Sine wave	20.	360
21.	180	22.	Positive
23.	Negative	24.	Frequency

Unit 1 Examination

Basics of Alternating Current (AC)

Instructions: For each of the following questions, circle the answer that most correctly completes the statement.

1. What is the approximate peak-to-peak voltage of an ac voltage of 6 V RMS?
   a. 15 V
   b. 17 V
   c. 19 V
   d. 21 V

2. The value of ac that would have the same effect in power produced as a similar value of dc is known as
   a. Peak value
   b. RMS value
   c. Average value
   d. Peak-to-peak value

3. What is the approximate peak-to-peak voltage of an ac voltage of 2 V RMS?
   a. About 6 V
   b. About 7 V
   c. About 8 V
   d. About 9 V

4. The waveform illustrated in Fig. E-3 represents
   a. Steady dc
   b. Pulsating dc
   c. Nonsinusoidal ac
   d. Sinusoidal ac

5. What is the peak voltage, if the effective value is 120 V?
   a. 169 V
   b. 338 V
   c. 84 V
   d. 90 V

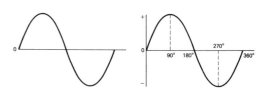

Fig. E-3.

6. What is the peak-to-peak voltage of an ac voltage equal to 10 V effective value?
   a. 2.82 V
   b. 14.1 V
   c. 28.2 V
   d. 141 V

7. If a wave is found to have a frequency of 250 Hz, its period is
   a. ¼ s
   b. 1/25 s
   c. 1/250 s
   d. 1/400 s

8. The *maximum* value of an ac voltage is called
   a. Instantaneous
   b. Effective
   c. Average
   d. Peak

9. The effective voltage of an ac waveform the peak value of which is 340 V is

    a. 170 V b. 240 V
    c. 300 V d. 340 V

10. The period of a sine wave is the time required for the sine wave to complete

    a. 1 cycle b. ½ cycle
    c. ¼ cycle d. ¾ cycle

11. If a sine wave of ac voltage contains 220 reversals of polarity per second, what is the frequency of the sine wave?

    a. 55 Hz b. 110 Hz
    c. 220 Hz d. 440 Hz

12. What is the approximate peak voltage of a 120 V ac wave?

    a. 85 V b. 120 V
    c. 169 V d. 240 V

13. The voltage generated through electromagnetic induction

    a. Can occur only when the conductor is wound in the form of a coil
    b. Cannot occur in a dc circuit
    c. Can be produced only when the coil is stationary and the magnetic field moves
    d. Is produced by moving a conductor through a magnetic field

14. An advantage of three-phase power over single-phase power is

    a. Smoother power output
    b. Three-phase motors have simple construction features
    c. Fewer conductors per kilowatt are required for distribution
    d. All the above

15. The line voltage of a delta-connected transformer is equal to

    a. 0 V
    b. Phase voltage
    c. Phase voltage × 1.73
    d. 58.8 V for 100 V operation

Basics of Alternating Current 13

True or false: Place either T or F in each blank.

_____ 16. If a three-phase ac generator has delta-connected windings, phase voltage will be equal to line voltage.

_____ 17. In a wye-connected three-phase ac generator, the line voltage is equal to the phase voltage multiplied by 1.73.

_____ 18. The electric degree separation of three-phase voltage is 120°.

_____ 19. Electric current is produced when a conductor moves through a magnetic field.

_____ 20. One cycle of ac voltage has a positive alternation and two negative alternations.

UNIT **2**

Measuring AC

An important activity in the study of electronics is measurement. Measurements are made in many types of electronic circuits. The proper ways of measuring ac should be learned. Common methods of measuring ac include the use of multimeters (VOMs) and oscilloscopes. Use of a multimeter is discussed in *Understanding DC Circuits*. This unit stresses the use and operation of oscilloscopes.

> *UNIT OBJECTIVES*
>
> Upon completion of this unit, you should be able to:
>
> 1. Measure alternating current with an ac ammeter.
> 2. Use a multimeter to measure ac voltage.
> 3. Explain the basic operation of an oscilloscope.
> 4. Demonstrate how an oscilloscope can be used to measure the amplitude, period, frequency, and phase relations of ac waveforms.
> 5. Perform ac circuit measurements and calculations.

Important Terms

Before studying unit 2, you should know the following terms:

Attenuation The reduction of a value.

Axis A straight line about which a body or geometric figure rotates or may be supposed to rotate.

Cathode ray tube (CRT) A vacuum tube that emits electrons from a cathode that are formed into a narrow beam and accelerated toward a phosphorescent screen. A visible pattern of light energy is produced when the electron beam strikes the screen.

Deflection Horizontal and vertical electron beam movement in the operation of a CRT.

Electron beam A narrow stream of electrons released from the gun electrodes of a CRT.

Electron gun Tube electrodes responsible for forming a narrow beam of electrons in the operation of a CRT.

Horizontal axis The sweep plane of a CRT that is parallel to the earth's horizon.

Oscilloscope (scope) An electronic instrument that has a CRT to display graphic representations of electric waveforms.

Probe The pointed tip of an instrument that makes contact with the electric circuit being examined.

Retrace The process by which an electron beam returns to its original starting position.

Sawtooth wave A waveform characterized by a slow, linear rise time and a virtually instantaneous fall time. This wave resembles the teeth of a saw.

Sweep In a CRT, the horizontal movement of the electron beam.

Synchronization (sync) The process of keeping vertical and horizontal signals of an oscilloscope in step with each other.

Time base A voltage generated by the sweep circuit of a CRT so that its trace is linear with respect to time.

Trace The pattern or fine-line display produced by the electron beam of a CRT.

Triggering To cause by means of one circuit the action of another circuit to start its operation. The horizontal sweep of an oscilloscope may be triggered into operation externally, by the power line, or by a vertical signal.

Vertical axis The deflection plane of a CRT that causes the electron beam to move up and down.

Measuring AC Voltage with a Multimeter

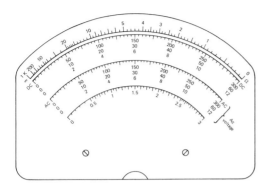

Fig. 2-1. Scale of an analog multimeter with ac voltage section.

A multimeter (VOM) may be used to measure ac voltage. AC voltage is measured in the same way as direct current (dc) voltage with the following two exceptions:

1. In the measurement of ac voltage, proper polarity does not have to be observed.
2. In the measurement of ac voltage, the ac voltage ranges and scales of the meter must be used.

Figure 2-1 shows the scale of an analog multimeter. The portion used to measure ac voltage is marked. The scales used to measure ac voltage are similar to those used to measure dc voltage.

Measuring AC Voltage with an Oscilloscope

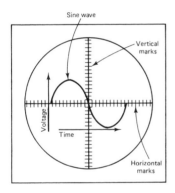

Fig. 2-2. AC waveform displayed on the screen of an oscilloscope.

Another way to measure ac voltage is with an oscilloscope. Oscilloscopes, or "scopes," are used to measure a wide range of frequencies with precision and to examine wave shapes. For electronic servicing, it is necessary to be able to see the voltage waveform while troubleshooting.

When the controls are properly adjusted, an oscilloscope allows various voltage waveforms to be analyzed visually with an image on a screen. This image, called a *trace,* is usually a line on the screen or cathode ray tube (CRT). A stream of electrons striking the phosphorescent coating on the inside of the screen causes the screen to produce light.

An oscilloscope displays voltage waveforms on two axes, as on a graph. The horizontal axis on the screen is the *time axis.* The vertical axis is the *voltage axis.* An ac waveform is displayed on the CRT as shown in Fig. 2-2. For the CRT to display a trace properly, the internal circuits of the scope must be properly adjusted. These adjustments are made with controls on the front of the oscilloscope. Oscilloscopes vary, but most have some of the following controls:

1. *Intensity:* Controls the brightness of the trace and sometimes is the on-off control.
2. *Focus:* Adjusts the thickness of the trace so that it is clear and sharp.
3. *Vertical position:* Adjusts the entire trace up or down.
4. *Horizontal position:* Adjusts the entire trace left or right
5. *Vertical gain:* Controls the height of the trace.

Measuring AC 17

6. *Horizontal gain:* Controls the horizontal size of the trace.
7. *Vertical attenuation, or variable volts/centimeter:* Acts as a coarse adjustment to reduce the trace vertically.
8. *Horizontal sweep, or variable time/centimeter:* Controls the speed at which the trace moves across (sweeps) the CRT horizontally. This control determines the number of waveforms displayed on the screen.
9. *Synchronization select:* Controls how the input to the scope is locked in with the circuitry of the scope.
10. *Vertical input:* External connections used to apply the input to the vertical circuits of the scope.
11. *Horizontal input:* External connections used to apply an input to the horizontal circuits of the scope.

The following procedure is used to adjust the controls of an oscilloscope to measure ac voltage. The names of some of the controls vary on different types of oscilloscopes.

1. Turn on the oscilloscope and adjust the intensity and focus controls until a bright, narrow, straight-line trace appears on the screen. Use the horizontal position and vertical position controls to position the trace in the center of the screen. Adjust the horizontal gain and variable time/centimeter until the trace extends from the left side of the screen to the right side of the screen. This allows display of the entire waveform.
2. Connect the proper test probes into the vertical input connections of the oscilloscope.
3. Measure the ac voltage.
4. After a waveform is displayed, adjust the vertical attenuation (volts/centimeter) and vertical gain controls until the height of the trace equals about 2 inches or 4 cm. Most scopes have scales that are marked in centimeters (cm). Adjust the vernier or stability control until the trace becomes stable. One ac waveform or more should appear on the screen of the scope.

Oscilloscope Operation

An oscilloscope is an instrument designed to measure time-varying voltage and current values graphically. This method of measurement allows the operator actually to see voltage and

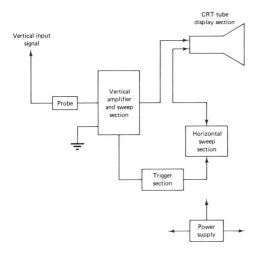

Fig. 2-3. Functional parts of an oscilloscope.

current signal traces rather than viewing the results on a deflection meter or a digital display instrument. In making measurements of this type, the oscilloscope takes very little energy away from the circuit being analyzed. It also responds well to irregularly shaped signal voltages, high frequency, and phase relationships. An oscilloscope is used for calibration of instruments, evaluation of the performance of equipment, and troubleshooting.

The primary function of an oscilloscope is to measure time-varying signals and display these signals so that they can be analyzed. As a rule, a user must prepare this instrument for operation before it can be used to make a suitable display. The setup procedure usually is easy, but it is helpful to know something about basic oscilloscope control functions.

An oscilloscope may be viewed as a series of functional blocks connected together to form an operating system. Presentation of the block structure is an organization procedure that allows the system to be viewed as a functional instrument. It is important to understand the specific role of each block in the operational system and the controls and setup procedures needed to make the oscilloscope function.

The fundamental blocks or parts of an oscilloscope connect together and cause it to respond as an operational system. In general terms, the system is composed of a display device, a horizontal sweep section, vertical amplification and sweep, triggering, power supply, and probes. These functional parts are the same for nearly all oscilloscopes. Figure 2-3 shows these parts in a block diagram.

The blocks of a basic oscilloscope are named according to the specific function they achieve in the operation of the instrument. The *vertical section*, for example, controls the Y axis, or vertical part of the display. The vertical section controls the up or down motion of the electron beam. The *horizontal section* of the basic instrument controls left-to-right movement of the electron beam. This part of the instrument produces the X axis of the display. The *trigger section* determines the specific point in time at which horizontal sweep begins. Triggering is achieved by means of switching action. The *CRT* is ultimately responsible for graphic display of the signal, the end result of all parts of the system working together. The *power supply* develops all the operational voltages needed to energize the circuit components. The *probe* serves as an external input receptacle for the instrument. The combined actions of these functions are needed to make the oscilloscope operational.

CRT

The CRT is responsible for the display function of an oscilloscope. Structurally the CRT is a long, evacuated glass tube in which electrons produced at the neck end of the device cause an image to appear on a glass surface in the display area. Electrons are produced, accelerated, and focused in the rear assembly of the tube. This part of the tube is called the *electron gun*, or simply the gun. Horizontal and vertical deflection plates are located near the neck area. The electron beam passes through these plates as it moves toward the display area. The large-diameter end of the tube is the display, or viewing, area. The inside glass face of the viewing area is coated with a phosphorescent material. When the high-velocity electron beam strikes this material, it produces a characteristic glow. The CRT changes an invisible electron beam into light energy that is displayed on a glass surface.

Figure 2-4 shows the construction details of a CRT. Figure 2-5 illustrates the operation of the electron gun assembly in the neck of the CRT. A beam of electrons is produced by the cathode of the tube when it is heated by means of application of a filament voltage. Electrons emitted from the cathode are initially attracted by the positive potential of anode 1. The quantity of electrons passing toward anode 1 is determined by the amount of negative bias voltage applied to the control grid. Anode 2 is operated at a higher positive potential than anode 1 to accelerate the electron beam further toward the screen of the CRT. High voltage in the display area of the tube serves as the final attracting force for the electron beam. As it strikes the phosphorescent screen, light is produced.

Control of the electron beam is achieved by means of a number of different processes. The quantity of electrons that reach the display area of the CRT determines brightness or intensity level. The intensity control, usually connected to the grid, determines the negative voltage value of the grid. The grid is a small, cylinder-like structure; the end nearest the cathode is open and the other end has a small aperture. The number of electrons forced to pass through the aperture depends on the amount of negative voltage on the grid. High negative voltage repels large numbers of electrons and reduces the level of intensity. Reduced negative voltage increases the quantity of electrons that reach the display area and increases intensity. The intensity control usually is located on the front panel of the oscilloscope.

The sharpness of the display image or trace is determined with the *focus* control. In a CRT, focus is controlled electrostatically. Electrons emitted from the cathode have a natural tendency to separate, or spread apart, as they move toward the face of the CRT. Each electron, because it has a negative charge, is repelled by its neighboring electrons. This action would normally cause the trace to appear as a large fuzzy ball on the face of the CRT. To alter this condition, focus is achieved by means of altering the difference in the positive voltage between anodes 1 and 2. As a rule, anode 1 is varied and anode 2 remains at a fixed value. Variations in positive voltage cause the trajectory angle of the electron beam to change by means of altering the shape of the electrostatic field. As a result of this adjustment, one can alter the point of convergence of the electron beam by changing a voltage value within the tube. The focus control usually is located on the front panel of the oscilloscope. In most oscilloscopes the focus and intensity control adjustments are interrelated. Adjustment of one control usually necessitates adjustment of the other control.

Deflection

The position of an electron beam in the display area of a CRT is controlled by two sets of deflection plates. These plates are housed in the neck of the CRT and are located between the electron gun and the display area. Figure 2-5 shows the position of

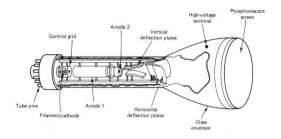

Fig. 2-4. CRT construction.

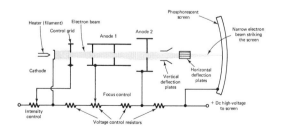

Fig. 2-5. Electron gun assembly.

these deflection plates near the center of the CRT. The set of plates closest to the gun assembly controls vertical deflection. The second set of plates controls horizontal deflection. The combined effect of these two sets of plates causes the electron beam to have both vertical and horizontal deflection at the same time.

Electron beam deflection of a CRT is accomplished by means of electrostatic charge energy. This energy is produced by means of applying voltages of the correct polarity to the deflection plates. Vertical deflection voltages normally are derived from an external source of energy. This generally represents the signal being viewed on the face of the CRT. Horizontal sweep voltages are developed internally by a time-based generator. Most oscilloscopes have optional circuitry that allows the horizontal sweep signal to be developed externally and applied to the deflection plates. External sweep is selected with a switch.

An end view of a CRT observed from the display area is shown in Fig. 2-6. This representation shows the location of the vertical and horizontal deflection plates with respect to the electron beam. If no voltage is applied to the deflection plates, the electron beam positions itself in the center of the display area. To demonstrate how deflection is achieved, we examine how the electron beam responds when voltage is applied to the deflection plates. We describe the response when voltage is applied to only one set of plates and discuss the response of the electron beam when sweep voltage is applied to both sets of plates at the same time.

Figure 2-7 shows electron-beam response when voltage is applied only to the vertical deflection plates. If the top vertical plate is made positive and the bottom plate negative, the electron beam is attracted by the top plate and repelled by the bottom plate (Fig. 2-7a). Reversing this polarity causes the beam to move toward the bottom of the CRT (Fig. 2-7b). AC voltage applied to the two plates (Fig. 2-7c) causes the electron beam to sweep from top to bottom according to the applied frequency. High-frequency ac causes the electron beam to appear as a solid vertical line. Low-frequency ac causes a slow-moving dot to be produced by the electron beam. The resulting dot produced by the electron beam moves slowly between the top and bottom of the CRT. The amount of voltage applied determines the length of the vertical sweep pattern.

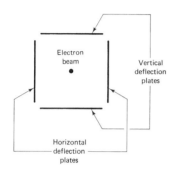

Fig. 2-6. End view of deflection plates.

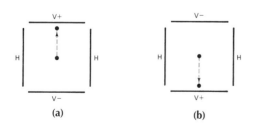
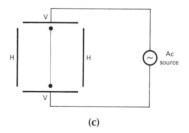

Fig. 2-7. Vertical deflection response. (a) Positive voltage applied. (b) Positive voltage applied to lower plate. (c) AC voltage applied between plates.

Measuring AC 21

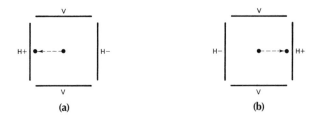
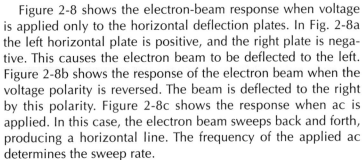

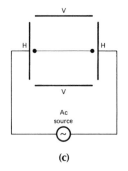

Figure 2-8 shows the electron-beam response when voltage is applied only to the horizontal deflection plates. In Fig. 2-8a the left horizontal plate is positive, and the right plate is negative. This causes the electron beam to be deflected to the left. Figure 2-8b shows the response of the electron beam when the voltage polarity is reversed. The beam is deflected to the right by this polarity. Figure 2-8c shows the response when ac is applied. In this case, the electron beam sweeps back and forth, producing a horizontal line. The frequency of the applied ac determines the sweep rate.

Fig. 2-8. Horizontal deflection response. (a) Positive voltage applied to left plate. (b) Positive voltage applied to right plate. (c) AC voltage applied to horizontal plates.

Internally generated horizontal sweep signals generally have a *sawtooth* shape. This signal, shown in Fig. 2-9, has a linear rise time and a rapid fall time. The rising portion of the wave is called the *ramp,* and the falling part of the wave is called the *retrace.* The time or space after retrace is called the *holdoff* area. The ramp portion of the wave causes sweep from left to right, and retrace causes return of the electron beam to the left side of the display. The holdoff part of the wave determines how long the trace must wait between trigger pulses. The frequency of the sawtooth wave is determined by the sweep rate of the internal horizontal time-base generator.

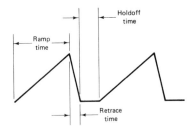

Fig. 2-9. Sawtooth waveform of the horizontal time-base generator.

When they are applied simultaneously to the horizontal and vertical deflection plates, sweep signals cause a specific pattern to appear on the face of the CRT. Assume now that an unknown sine wave is to be viewed on the oscilloscope. This wave is supplied to the vertical input for application to the deflection plates. An oscilloscope usually is set up for internal horizontal sweep-signal generation. A sawtooth wave is generated by the time-base generator and applied to the horizontal deflection plates. Figure 2-10 shows the display pattern that appears on the CRT as a result of the applied signals. The beam sweeps from left to right according to the horizontal signal. In the same time frame, the vertical signal causes the electron beam to be deflected up and down according to its voltage value. At any specific point in time, the position of the electron beam is determined by the combined forces of the vertical and horizontal deflection voltage values. These forces cause the sine wave of the vertical input signal to be produced on the CRT. The resulting display is four complete sine waves.

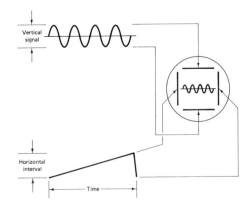

Fig. 2-10. Time-varying signals simultaneously applied to horizontal and vertical deflection plates.

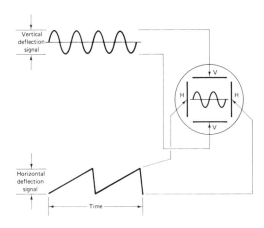

Fig. 2-11. Display of Fig. 2-10 with horizontal deflection frequency doubled.

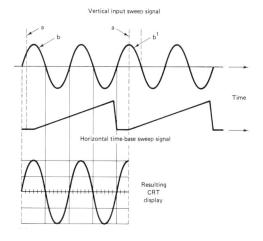

Fig. 2-12. Synchronized vertical and horizontal sweep signals and resulting CRT display.

Changing the time period of the horizontal sweep voltage alters the display on the CRT. In Fig. 2-11 the horizontal sweep frequency has been doubled without a change in the vertical input signal. In this case, doubling the horizontal frequency causes the time base to be halved. Time is a function of frequency and is expressed as $T = 1/f$. When the horizontal sweep is reduced to half its previous value, fewer sine waves are displayed on the CRT. This means that the frequency of the horizontal time-base generator determines the scale of the time axis. The scale or operational range of the time-base generator is controlled by a switch on the front panel of the oscilloscope. This switch is calibrated in units of time per division. Typical ranges are seconds/centimeter (s/cm), milliseconds/centimeter (ms/cm), or microseconds/centimeter (μs/cm). Specific values in an operating range might be 0.5, 0.2, 0.1 s/division; 50, 20, 10, 5, 2, 1, 0.5, 0.2, 0.1 ms/division; or 50, 20, 10, 5, 2, 1, 0.5, 0.2, 0.1, .05 μs/division.

Triggering and Synchronization

To produce a steady waveform on the face of the CRT, an oscilloscope must repeat the same trace path. For this to be achieved, the displayed signal must always begin its trace at the same point on the wave. Another way of describing this function is to say that the start of the sawtooth voltage must be synchronized with a specific point on the displayed signal. Synchronization of the vertical and horizontal signals is a function of the trigger system.

Figure 2-12 shows signals that are applied to the vertical and horizontal deflection systems and the resulting display produced by an oscilloscope. The top display is the waveform supplied to the vertical deflection system. The middle waveform is produced by the time-base generator and applied to the horizontal deflection system. The bottom trace is the resulting waveform displayed on the CRT. The time-base signal starts its sweep at point a of the sine wave. When the time-base signal completes one trace period, it drops back to its initial voltage value and waits for the next sine wave to return to its beginning point before starting the next trace. This is indicated as point a on the display. Because the resulting image is of a periodic nature and the time base is synchronized to produce exactly the same portion of the signal on each trace, the display image appears to be a stationary sine wave. If the applied signal voltage were to change during the sweep period or if the sweep began at a random point on the input signal voltage, the resulting image would be different for each trace period. This would make the display unsteady or appear to be in a continuous state of motion.

Two controls on an oscilloscope allow the operator to select the point on the display signal where sweep begins. The trigger-level control sets the voltage value that causes the sweep to start. This control alters the starting point or beginning of the display. This function usually is achieved by means of a variable control adjustment. The slope control adjusts the polarity of the voltage where sweep begins. On the display in Fig. 2-12, triggering occurs on the positive slope. If the negative slope were used as the trigger point, sweep would occur at both b points. This would cause the vertical part of the trace to start with a downward motion instead of an upward motion. The slope function is achieved by changing a switch. The slope is either positive or negative according to the desires of the operator.

An oscilloscope generally has some additional trigger controls, depending on its design. One of these is described as the triggering mode of operation. Trigger-mode selection is achieved with a switch. If the EXT trigger mode is selected, the oscilloscope is triggered by an external signal. This signal is derived from the circuit under test or from an external generating source that is used as a reference. With external triggering, it would be possible to observe the phase relations of amplifier input and output signals. Selection of the LINE trigger mode allows the time-base generator to be synchronized by the ac line voltage. This allows the oscilloscope to detect the source of unwanted signals. Noise on a waveform can be made stationary when the signal is synchronized with the line voltage.

Selection of the AUTO mode switch places the oscilloscope in the automatic-triggering mode of operation. Triggering in this manner is considered to be the automatic signal-seeking mode of operation. Assume that a trigger pulse starts the sweep signal. This action causes the electron beam to sweep during the ramp time and retrace period until the holdoff period ends. At this point a timer begins to run. If another trigger does not occur before the timer runs out, a pulse is generated automatically. This allows most signals to be displayed automatically because triggering is duplicated by the internal circuitry of the oscilloscope. This mode of operation also allows the operator to trigger on signals with changing voltage amplitudes or shapes without making level-control adjustments. Automatic triggering is probably the most widely used mode of operation for general oscilloscope work.

Power Supply

The power supply of an oscilloscope is primarily responsible for supplying dc and in some cases ac voltage to the active and passive components of the instrument. These voltage values are derived from the ac power line. Typical primary-line voltage is 120 V at 60 Hz. This voltage is applied to a transformer, which steps the voltage up or down according to the needs of the circuit. The voltage values developed by the supply depend on the circuitry of the oscilloscope and the components being used. For the most part, modern oscilloscopes use solid-state devices. Bipolar transistors, junction field-effect transistors (JFETs), and integrated circuit devices are used in the circuits. As a rule, these devices respond to low-voltage dc energy. The CRT of an oscilloscope, however, requires some rather sizable dc voltage values. DC voltages of 100 to 2000 V are needed to energize the electrodes. The anode of the CRT also may necessitate a dc voltage in the range of 5 to 10 kilovolts (kV). This part of the power supply requires some rather unusual circuitry, such as voltage multipliers or a special high-voltage transformer to develop the necessary voltage values.

The high-voltage supply of an oscilloscope is generally described as a *generated voltage source*. It develops high-voltage ac from a low-voltage dc source. The dc energizes a transistor that responds as an oscillator. High-frequency ac is supplied to the primary winding of the transformer. The transformer of this circuit can be made smaller because of the high-frequency ac.

Oscilloscope Probes

The probe of an oscilloscope is responsible for connecting voltage and current signals to the vertical input terminals without loading or disturbing the circuit under test. To meet these requirements, a variety of probes are available, from simple passive units to sophisticated active probes for special measurement applications. In each case, the probe must not degrade the performance of the oscilloscope, and it should be properly calibrated to ensure measurement accuracy.

Figure 2-13a shows a general block diagram of an oscilloscope probe. The probe head contains the signal-sensing device. In a passive probe this is achieved with a 10 megohm (MΩ) resistor shunted by a 7 picofarad (pF) capacitor. Active probes have a bipolar transistor, or JFET device, in the probe head. Coaxial cable is used to couple the probe head to the termination circuitry. The termination provides the oscilloscope with the source impedance needed to connect the coaxial cable to the vertical input circuitry.

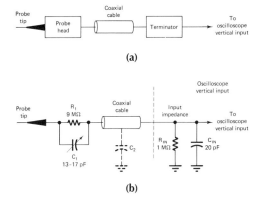

Fig. 2-13. Oscilloscope probe diagrams.

Passive probes are widely used in most oscilloscope applications. The simplest passive probe is a nonattenuating, or ×1, unit. This type of assembly consists simply of a length of coaxial cable with a probe tip at one end and a cable connector at the other end. Connection is achieved directly through the coaxial cable. Electrically the coaxial cable has some shunting capacitance that must be taken into account. A 50 ohm (Ω) coaxial cable has a shunting capacitance of 30 pF/ft. A 5 ft length of coaxial cable would therefore offer approximately 150 pF of shunt capacitance to the oscilloscope input. This type of probe offers a rather large shunt capacitance to high-frequency ac. As a rule ×1, or direct, probes are used exclusively to measure low-frequency and power-line circuits.

One of the most widely used passive probes is the compensating ×10 attenuating unit (Fig. 2-13b). This probe provides signal attenuation in a ratio of 10:1 over a wide range of frequencies. The probe head has a 9 MΩ resistor shunted by a small variable capacitor. Adjustment of the variable capacitor allows the probe to compensate for impedance changes over a wide range of frequencies. An applied signal should be attenuated by a factor of 10 but still maintain its original shape and phase without distortion. Compensation adjustments allow the probe to couple sample signals to the vertical input without producing distortion and adversely loading down the circuit being evaluated. This adjustment is achieved by means of applying a representative square-wave signal of the frequency being analyzed with the probe. The compensating capacitor is adjusted to produce the best reproduction of the square wave on the CRT. Compensation can be made with a screwdriver adjustment or by means of twisting a barrel capacitor over the probe tip. The structure of the variable capacitor of the probe usually dictates the adjustment procedure.

Oscilloscope Controls

The operating controls of an oscilloscope usually are divided into convenient groups that alter or control a specific instrument function. These include CRT control, vertical deflection, horizontal sweep, triggering, modes of operation, and probes. Each group has a number of unique controls in its makeup. In general, these controls are designed to perform some type of operational procedure. Operational controls are usually placed in a convenient location where they can be easily adjusted.

CRT Display Controls. The CRT display group of oscilloscope controls consists of intensity, focus, trace rotation, and beam finder. These controls are generally located in a position relatively close to the viewing area of the CRT.

Intensity Control. The intensity control of an oscilloscope is used to adjust the brightness level of the display. This is achieved by means of altering the amount of negative voltage going to the control grid of the CRT. Intensity of the electron beam is determined by the quantity or number of electrons that reach the viewing area. When the negative voltage of the grid is reduced in value, a larger number of electrons reach the display area. Increased negative voltage reduces the brightness or intensity of the electron beam. In practice, the intensity control should be adjusted to produce the lowest level of brightness that will allow the display to be viewed effectively. The operational life of the CRT can be prolonged when the intensity of the electron beam is kept to a minimum. Functionally, this control adjusts the brightness level of the trace so that it can be viewed in different ambient light conditions.

Focus. The focus control of an oscilloscope is used primarily to alter the sharpness of the display image. Focus is achieved by means of altering the voltage level of the first anode of the CRT. The potential difference in voltage between first and second anodes of the CRT determines the strength of an electrostatic field. This condition alters the trajectory of the electron beam. A rather large difference in voltage causes the beam to converge into a fine trace near the surface of the viewing area. Focus is simply a voltage adjustment that allows the trace to have fine detail. In most instruments, focus and intensity adjustments are interrelated.

Trace Rotation. The trace rotation control of an oscilloscope is another one of the CRT group of controls included on the front panel of the instrument. Trace rotation allows the user to align the horizontal trace of the display electrically with fixed lines on the graticule (the grid pattern). This control is generally less accessible than the other controls. This is intentional to prevent accidental misalignment of the control. In most oscilloscopes, this adjustment is made with a small screwdriver. Once the adjustment has been made, it generally does not have to be made again unless the instrument is subjected to a stray magnetic field or moved to a different location. In portable instruments this control is very handy because of the different locations in which the instrument is used.

To adjust trace rotation, turn on the instrument and make the necessary adjustments to allow a single horizontal line to be displayed across the center of the display. Position the line so that it is aligned with one of the horizontal lines of the graticule. Adjust the control so that the trace line is parallel to the graticule lines. It may be necessary to adjust the vertical position of the trace again to assure that the two lines are parallel.

Beam Finder. The beam finder of an oscilloscope is a convenience control that allows the user to locate the electron beam when it is off screen. The beam finder is a push-button switch usually located in the CRT group of controls. When this button is depressed, the vertical and horizontal deflection voltages are reduced. This allows the trace to be displayed in the limited space of the face of the CRT. When the operator sees the location of the beam, the vertical and horizontal position controls can be adjusted to center the trace. Some oscilloscopes do not have a beam-finding switch.

Vertical Deflection Controls. The vertical section of an oscilloscope supplies the display part of the instrument with vertical information that ultimately appears on the CRT. The vertical section takes the input signal, amplifies it, and develops a suitable voltage value that deflects the electron beam. The input signal usually is the signal or voltage being analyzed with the oscilloscope. Figure 2-14 is a block diagram of the vertical section of a typical oscilloscope. Controls are attached to different parts of the system. Typical controls are vertical position, input coupling, vertical operating mode, input sensitivity or volts/division selector, and variable volts/division control.

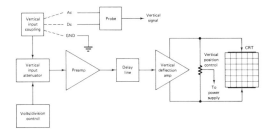

Fig. 2-14. Block diagram of the vertical sweep section.

Vertical Position Control. The vertical position control allows the operator to adjust the trace to a desired viewing location. This control adjusts the distribution of voltage between the two deflection plates. If the voltage is equally distributed between the deflection plates, the trace is centered vertically. Adjusting the control so that the top plate is more positive than the bottom plate moves the display toward the top of the CRT. Reversing the voltage distribution causes the trace to be positioned near the bottom of the display area.

Vertical Input Coupling. The input coupling function of an oscilloscope is controlled with a switch. This function lets the operator control how the input signal is coupled with the vertical input section of the instrument. In the ac position, the input signal voltage must pass through a capacitor. In the dc position, the input signal is coupled directly to the input of the vertical amplifier. The middle position, or GND, refers to the ground. Placing the switch in this position disconnects the external input signal from the vertical section and grounds the vertical input.

The influence of control on the operation of an oscilloscope can be seen when an appropriate signal is applied to the vertical input. An *ac signal* adjusts its zero reference point to a value determined by the vertical position control. A *dc signal* adjusts its reference to a level determined by the voltage value of the applied dc. The ground switch position selects the chassis ground as a reference operating point.

Volts/Division. The volts/division or vertical sensitivity of an oscilloscope is controlled with a rotary switch. This control allows the instrument to extend its range of operation so that signals of a few millivolts to several volts can be displayed on the CRT. Using the volts/division switch also changes the scale factor of the display screen. Each position setting of the switch has a number value that represents the scale factor. The 10 position means that each major vertical division represents 10 V. Other position settings cause a corresponding volts/division function to be established.

All these marked values depend on the variable control being set to the calibrate position. The total amount of vertical sweep or deflection is based on the *peak-to-peak voltage* of the ac signal being displayed. The probe of an oscilloscope also may have some influence on the amount of vertical sweep produced by the instrument. A ×1 probe provides a direct reading of the volts/division ranges, whereas a 10 probe scales down the input by a factor of 10. The probe and volts/division switch setting of the instrument determine the range of vertical sweep.

Variable Volts/Division Control. Most oscilloscopes have a variable volts/division control that can be used to change the volts/division range setting by a factor of ×2 or more. This control is used to make quick amplitude comparisons of signals. It changes the range of the volts/division setting. In most instruments this control has a notch or position setting where calibration is actuated. The amplitude of the display is reduced or increased with this adjustment. For most oscilloscope applications, the variable control should remain in the calibrate position.

Vertical Operating Mode. The vertical operating mode of an oscilloscope is an operational control for instruments that have two vertical channels. Some oscilloscopes have a duplicate set of vertical controls for each channel. Two independent traces can be displayed on the CRT at the same time. The vertical operating mode is usually a switch function that allows the operator to select a desired channel or a combination of channel options.

The vertical mode switch of a channel selects channel 1, channel 2, or both. The mode switch of a channel often has ADD, ALTERNATE, and CHOP modes of operation. For example, if the channel 1 mode switch is in the BOTH position and the channel 2 switch is in the alternate, or ALT, position, you can adjust the position controls so that the trace of channel 1 is at the top and the trace of channel 2 is near the bottom. If an ac signal is applied to each channel, the volts/division switch of each channel may be adjusted to produce a display of suitable size. The mode switch in the ADD position adds the two traces. The CHOP and ALT mode switches are used to observe two signals at any sweep speed. The alternate mode displays one channel and then the other in an alternate sequence. At high sweep rates, this type of display is very desirable. At slow sweep rates there is a noticeable alternating effect in the two displays. The CHOP mode breaks the two traces into small segments and switches between the two traces very quickly. This is generally noticeable when 60 Hz signals are displayed on the instrument.

Horizontal Sweep Controls. For an oscilloscope to make a display on the face of a CRT, it needs both vertical and horizontal sweeps to deflect the electron beam. Horizontal sweep is normally provided by an internal generator, which produces a sawtooth-shaped waveform. The rising part of this waveform is called the *ramp,* or trace, period, and the falling part is called the *retrace* interval. The trace period causes beam deflection from left to right. Retrace causes the beam to return to the left in preparation for the next trace period. The horizontal sweep rate of an oscilloscope is operator controlled, which allows it to display different frequencies.

Figure 2-15 is a block diagram of the horizontal sweep circuitry of an oscilloscope. The circuitry of an oscilloscope is divided into two sections. The generator is responsible for sweep-signal development. The amplifier increases the amplitude of the signal so that it will drive the deflection plates. Controls of this section are attached to the part of the circuit that has the greatest influence on its operation.

The horizontal sweep of an oscilloscope has a variety of different controls that regulate its operation. These include horizontal position, operational mode, seconds/division, variable sweep, and magnification. These controls are generally grouped together in a special area of the control panel.

Horizontal Position Control. The horizontal position control is designed to change the location of the horizontal trace on the face of the CRT. This adjustment is made by means of shifting the distribution of voltage to the horizontal deflection plates. The trace shifts in the direction of the deflection plate with the highest positive voltage value. As a rule, the horizontal trace should be positioned in the center of the viewing area. The position control should cause the trace to shift from right to left according to the setting of the control. If a dual-channel oscilloscope is used, the position control alters both channels in the same manner.

Horizontal Operating Mode. The horizontal operating mode is an optional function that depends on the degree of sophistication of the oscilloscope. Single time-base oscilloscopes usually have only one mode of operation. Some oscilloscopes are equipped with normal, intensified, and delayed sweep. As a rule, the instrument is used in the normal mode of operation for most applications. In this mode, the horizontal time base responds as an energy source for the horizontal sweep system.

The intensified mode of operation allows the operator to alter the electron beam with a signal voltage that causes its intensity to vary according to an external signal. This mode of operation

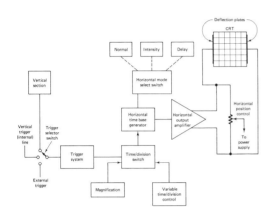

Fig. 2-15. Horizontal sweep section.

allows some rather unusual tests to be performed with the oscilloscope. The electron beam of the CRT of a television receiver is intensity modulated to produce light variations in the display.

Delayed sweep allows the instrument to add a precise amount of time between the trigger point and the beginning of the sweep period. With delayed sweep the operator may choose to trigger the trace anywhere along the displayed waveform. The delay time is used to control the start or beginning of the waveform being displayed. For normal measurement applications, the mode switch is placed in the no-delay mode of operation.

Horizontal Time-Base Control. The horizontal time base of an oscilloscope is used to generate a sawtooth wave that deflects the trace horizontally. This action is achieved with a switch that controls the sweep rate of the time-base generator. The switch positions are identified in time/division values. Three ranges of sweep are generally included in an oscilloscope, seconds/division, milliseconds/division, and microseconds/division. A number of discrete values are included in each of these ranges.

The function of this switch is to select the sweep frequency of the horizontal time base. When the frequency of the time base coincides with the frequency of the signal being applied to the vertical input, a suitable display is produced. If the time-base rate is greater or less than the frequency of the observed signal, an unintelligible signal is displayed. The frequency of the time-base generator must be adjusted to a reasonable approximation of the observed signal frequency to produce a meaningful display. Operation of this control is simply a matter of selecting an appropriate sweep range to produce a usable display.

Variable Time/Division Control. The switch positions of the time/division control are designed to provide calibrated time-base values. Sometimes there is a need for variable control of the time base. The variable time/division control allows the operator to adjust the time base. This control is located at the center of the time/division switch. The extreme clockwise position usually places the variable control in the calibrate position. When the control is out of the clockwise position, its variable condition is in effect. This control is left in the calibrate position for most measurement applications.

Horizontal Magnification. Most oscilloscopes offer some means of horizontally magnifying the waveforms that appear on the screen. Magnification is achieved by means of multiplying the time base by a fixed factor. A factor of 10 is typical. This is sometimes achieved by means of pulling out the variable time/division control. This action changes the time-base components by a factor of 10. A 0.05 μs signal can be extended to a 5 nanosecond (ns) signal by engaging the magnification switch. Magnification is very useful for analyzing specific parts of a signal.

Triggering Control. The time-base generator of the horizontal sweep system is considered to be free running. The frequency of the generated sawtooth waveform is based on the *RC* time constant of circuit components. For this signal to be displayed, there must be some form of *synchronization*. The horizontal sweep signal must be in step with the vertical signal. The triggering function of an oscilloscope is responsible for selecting a synchronizing signal and applying it to the horizontal sweep generator. Most oscilloscopes are equipped with internal, line, and external triggering capabilities. The trigger system simply tells the oscilloscope which trigger source to use, according to its switch selection. The trigger is adjusted with the slope and level controls to recognize a particular voltage level and polarity. Typical controls of a trigger system are level, slope, variable holdoff, source selection, trigger mode, and coupling.

Variable Holdoff Control. Not every triggering event can be accepted as a trigger pulse for the time-base generator. The trigger system does not recognize triggering during the trace time, the retrace period, or the short time that follows retrace, called the *holdoff time*. The holdoff period does, however, provide additional time after the retrace period to produce stability. In some applications, the holdoff time may not be long enough to provide good stability of the display. The variable holdoff control allows adjustment of the holdoff time. Changing the holdoff time makes it possible for the instrument to accommodate a trigger point that will appear at the same position on the wave for each repetition of the signal.

Trigger-Operating Modes. The trigger-operating mode of an oscilloscope is used to select different triggering methods for the time-base generator. For example, the MODE switch may have three positions—automatic, normal, and TV field. The normal trigger mode is the most useful. It accommodates the widest range of signals. This mode of operation does not allow a trace to be displayed on the CRT unless the time-base generator is triggered.

In the automatic mode a trigger pulse starts the trace, retrace, and holdoff sequence. At the completion of the sequence, a timer begins its operation. If another trigger pulse does not occur before the timer runs out, a pulse is generated to trigger the next sweep sequence. The automatic mode of triggering is considered to be the *signal-seeking* mode of operation. This means that for most measuring applications, the automatic mode matches the trigger level to the trigger-signal value. Trigger levels in the automatic mode do not require a value setting outside the signal range. This mode also lets the time-base generator trigger on signals that have changing amplitudes or waveshapes without making level adjustments.

When the trigger-operating mode switch is placed in the TV field position, the oscilloscope becomes useful in television-signal analysis. In this operating mode, the time-base generator triggers on TV fields at 100 μs/division and TV lines at 50 μs/division. This allows the instrument to display video signals, horizontal frequency, vertical, and color-burst signals with good synchronization.

Trigger-Signal Sources. The trigger-signal source of an oscilloscope is divided into three groups. These are described as internal, line, and external. The trigger source of an oscilloscope does not effectively alter operation of the trigger circuit. An internal trigger signal, however, means that the signal being displayed by the CRT also is used to trigger the time-base generator. The triggering source and its switching procedure vary a great deal among different oscilloscope makes and models.

The internal triggering source is enabled when a switch is placed in the INT position. In this position triggering can be from either the vertical channel or the vertical mode of operation. Triggering is determined by the vertical signal. In the VERT MODE position, the trigger source is selected for any of the vertical combinations such as channel A + B, A − B, chopped, A only, or B only. In a sense, this trigger-mode selection procedure is considered to be an automatic internal source-selection procedure.

The LINE source of triggering allows an alternative to the internal triggering source derived from the vertical input signal. Line triggering is very useful for analyzing signals that derive their energy from the ac power source. Line triggering is enabled by means of placing the source switch in the LINE position. This selects the trigger signal from a sample of the ac power line.

An alternative to internal triggering is *external triggering*. This triggering source comes from an externally supplied signal. External triggering usually gives the operator greater control over the display. To use this triggering source, the operator places the selector switch in the EXT position. The trigger signal must be supplied to the instrument from an outside source. External triggering is useful for analyzing digital signals. An operator might want to look at a long train of very similar pulses while triggering with a signal derived from a clock or another part of the circuit. The external trigger signal also can be used as a reference source in phase analysis of amplifier circuits.

Trigger Coupling. When an external trigger source is applied to an oscilloscope, it generally has signal coupling. The external trigger-coupling circuit can be dc, dc with attenuation, or ac. The dc coupling circuit allows application of both ac and dc signals to the external trigger source. The dc with attenuation input is used to accommodate signals with voltage values greater than those needed for normal signal input. This external trigger-coupling circuit divides the input by a factor of 10. A 100 V input signal divided by 10 equals 10 V. The ac coupling circuit blocks the dc components of the signal and couples only the ac component.

Summary

To use an oscilloscope effectively, the operator must understand the operating controls. These controls set up the instrument for measuring applications.

Self-Examination

Match the oscilloscope control with the proper function. If the control on your scope has a different name, place it in parentheses beside the control name listed.

_____ 1. Intensity		A.	Selects horizontal sweep frequency
_____ 2. Focus		B.	Provides up and down adjustment of beam
_____ 3. Vertical position		C.	Determines amplitude of horizontal deflection
_____ 4. Horizontal position		D.	Controls size of electron beam
_____ 5. Vertical gain		E.	Provides left and right adjustment of beam
_____ 6. Horizontal gain		F.	Controls brightness of beam
_____ 7. Vertical attenuation		G.	Eliminates vertical shift of display
_____ 8. Sweep select		H.	Connects voltages to the horizontal amplifier
_____ 9. Horizontal sweep		I.	Determines frequency of the sawtooth sweep
_____ 10. Synchronization select		J.	Determines amplitude of vertical deflection
_____ 11. Horizontal input		K.	Reduces vertical input amplitude
_____ 12. Vertical input		L.	Connects external signals to vertical amplifier
		M.	Determines amount of voltage used to synchronize sweep oscillator
		N.	Selects type of synchronization desired
		O.	Balances dc output
		P.	Sweeps at line frequency

Refer to Fig. 2-16 and determine each of the ac voltage values that correspond to the pointer location.

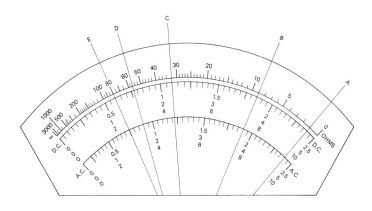

Fig. 2-16. VOM scale.

Measuring AC 33

AC voltage range	A	B	C	D	E
2.5 V ac	13. _____	14. _____	15. _____	16. _____	17. _____
10 V ac	18. _____	19. _____	20. _____	21. _____	22. _____
50 V ac	23. _____	24. _____	25. _____	26. _____	27. _____
250 V ac	28. _____	29. _____	30. _____	31. _____	32. _____
500 V ac	33. _____	34. _____	35. _____	36. _____	37. _____
1000 V ac	38. _____	39. _____	40. _____	41. _____	42. _____

Answers

1. F	2. D	3. B	4. E
5. J	6. C	7. K	8. A
9. I	10. N	11. H	12. L
13. 2.4	14. 1.9	15. 1.225	16. 0.9
17. 0.75	18. 9.6	19. 7.6	20. 4.9
21. 3.6	22. 3	23. 48	24. 38
25. 29	26. 18	27. 15	28. 240
29. 190	30. 122.5	31. 90	32. 75
33. 480	34. 380	35. 290	36. 180
37. 150	38. 960	39. 760	40. 490
41. 360	42. 300		

Experimental Activities for AC Electronics

The experimental activities that follow emphasize the practical applications of electronics. They parallel the content of each of the units in this book. The expense of the equipment is kept to a minimum. A few activities require no lab equipment. Each experimental activity is organized in the following way.

Experiment 2-1
(First activity of Unit 2)

Title
(Topic of the activity)

A few introductory paragraphs containing an overview of the activity, practical applications, the purpose of the activity, and suggested observations that should be made.

OBJECTIVE — Expected learning to take place through the experiment.

EQUIPMENT — Equipment and materials necessary to perform the experiment.

PROCEDURE — Logical, step-by-step sequence for completing the learning activity. Maximum use is made of charts and tables that aid in the recording of data.

ANALYSIS — Specific questions and problems that supplement the experimental activity.

The experimental material is presented in a single-concept approach. Activities organized in this way require only a short time to assemble and make the necessary measurements to facilitate learning.

In this book several experimental activities are used to reinforce the text material. These activities provide a different direction for the learning process. As a rule, the activity is experimentally based. This involves some manipulative activity, or hands-on operation. The activities deal with circuit construction, testing operations, calculations, instrument use, and identification and use of components. Through this approach you will become more familiar with electronic components and their use in a specific circuit application.

Tools and Equipment

A variety of tools and components are needed to perform the experimental activities in this course. These may be obtained from electronics supply houses, mail-order supply houses, and educational vendors. A listing of these sources appears in appendix C.

Important Information

At this time you may want to turn to the back of the book and review the following information:

Appendix A: Electronics Symbols

Appendix D: Soldering Techniques

The information in these sections will help you perform the experimental activities in this book.

Lab Activity Troubleshooting and Testing

The lab activities included in this book provide an opportunity to practice troubleshooting and testing for electronic circuits, devices, and systems. This section is a comprehensive list of troubleshooting and testing procedures that may be accomplished while performing each lab activity. Emphasis should be placed on *understanding circuit operation* and *understanding proper use of test equipment*. A technician who understands how the circuit, device, or system functions and knows how to use test equipment will find troubleshooting and testing relatively easy. This is true even for the simplest type of electronic circuit.

Competencies for Troubleshooting and Testing

Specific competencies should be developed during completion of this book. Upon completion of the activities presented in this book, you should be able to do the following.

Objectives

1. Outline basic troubleshooting procedures for locating specific trouble with devices and equipment.
2. Find the parts or circuits that are defective by using a "common sense" approach.
3. Test devices and circuits of electronic equipment using correct procedures.

Competency List

Unit 2: Measuring AC

Experiment 2-1—Measuring AC Voltage

1. Construct a simple ac circuit.
2. Recognize the characteristics of ac circuits.
3. Use a multimeter to measure ac voltage.
4. Compare dc voltage to ac RMS voltage by making voltage measurements.
5. Draw ac voltage waveforms.
6. Convert ac voltage values from a given unit to another including effective (RMS), peak, peak-to-peak, and average values.
7. Convert ac frequency to period (time) and vice versa.

Experiment 2-2—Measuring AC with an Oscilloscope

1. Become familiar with the controls of an oscilloscope.

2. Describe how the adjustment of each oscilloscope control affects the display on the screen of an oscilloscope.
3. Use a signal generator or function generator as an ac voltage source.

Unit 3: Resistance, Inductance, and Capacitance in AC Circuits

Experiment 3-1—Inductance and Inductive Reactance

1. Recognize the effects of ac on the inductive reactance of a coil.
2. Make measurements for an inductive ac circuit with a multimeter.
3. Determine the inductive reactance of a coil by making measurements with a multimeter.
4. Recognize the rating of an inductor (in henrys).

Experiment 3-2—Capacitance and Capacitive Reactance

1. Recognize the effects of ac on the capacitive reactance of a capacitor.
2. Make measurements for a capacitive ac circuit with a multimeter.
3. Determine the capacitive reactance of a capacitor by making measurements with a multimeter.
4. Recognize the working voltage (WVdc) and capacitance (μF or $\mu\mu$F) ratings of capacitors.

Experiment 3-3—Series *RL* Circuits

1. Recognize the characteristics of a series *RL* circuit.
2. Measure voltage and current values of a series *RL* circuit with a multimeter.
3. Compare measured and calculated values of voltage and current of a series *RL* circuit.
4. Calculate phase angle, impedance, and power factor of a series *RL* circuit by making measurements with a multimeter.

Experiment 3-4—Series *RC* Circuits

1. Recognize the characteristics of a series *RC* circuit.
2. Measure voltage and current values of a series *RC* circuit with a multimeter.
3. Compare measured and calculated values of voltage and current of a series *RC* circuit.
4. Calculate phase angle, impedance, and power factor of a series *RC* circuit by making measurements.

Experiment 3-5—Series *RLC* Circuits

1. Recognize the characteristics of a series *RLC* circuit.
2. Measure voltage and current values of a series *RLC* circuit with a multimeter.
3. Compare measured and calculated values of voltage and current of a series *RLC* circuit.
4. Calculate phase angle, impedance, and power factor of a series *RLC* circuit by making measurements.

Experiment 3-6—Parallel *RL* Circuits

1. Recognize the characteristics of a parallel *RL* circuit.
2. Measure voltage and current values of a parallel *RL* circuit with a multimeter.
3. Compare measured and calculated values of voltage and current of a parallel *RL* circuit.
4. Calculate phase angle, power factor, true power, apparent power, reactive power (VAR), admittance, conductance, susceptance, and impedance of a parallel *RL* circuit by making measurements.

Experiment 3-7—Parallel *RC* Circuits

1. Recognize the characteristics of a parallel *RC* circuit.
2. Measure voltage and current values of a parallel *RC* circuit with a multimeter.
3. Compare measured and calculated values of voltage and current of a parallel *RC* circuit.
4. Calculate phase angle, power factor, true power, apparent power, reactive power (VAR), admittance, conductance, susceptance, and impedance of a parallel *RC* circuit by making measurements.

Unit 4: Transformers

Experiment 4-1—Transformer Analysis

1. Determine the voltage ratio and current ratio of a transformer by making measurements with a multimeter.
2. Sketch a schematic of a transformer.
3. Make resistance measurements on the primary and secondary windings of a transformer.

Unit 5: Frequency-sensitive AC Circuits

Experiment 5-1—Low-pass Filter Circuits

1. Calculate the theoretical frequency response for a low-pass filter circuit.
2. Make measurements with an oscilloscope or multimeter to plot a frequency response curve for a low-pass filter circuit.
3. Use a signal generator or function generator as a variable frequency ac source.

Experiment 5-2—High-pass Filter Circuits

1. Calculate the theoretical frequency response for a high-pass filter circuit.
2. Make measurements with an oscilloscope or multimeter to plot a frequency response curve for a high-pass filter.

Experiment 5-3—Band-pass Filter Circuits

1. Calculate the theoretical frequency response for a band-pass filter circuit.
2. Make measurements with an oscilloscope or multimeter to plot a frequency response curve for a band-pass filter.

Experiment 5-4—Series Resonant Circuits

1. Recognize the characteristics of a series resonant circuit.
2. Measure voltage values of a series resonant circuit with a multimeter or oscilloscope.
3. Determine resonant frequency of a series ac circuit by using a multimeter or oscilloscope.
4. Calculate quality factor and bandwidth of a series resonant circuit by making measurements.
5. Compare measured and calculated values of resonant frequency of a series resonant circuit.

Experiment 5-5—Parallel Resonant Circuits

1. Recognize the characteristics of a parallel resonant circuit.
2. Measure current values of a parallel ac circuit with a multimeter.
3. Determine resonant frequency of a parallel resonant circuit by using a multimeter or oscilloscope.
4. Calculate impedance, quality factor, and bandwidth of a parallel resonant circuit by making measurements.
5. Compare measured and calculated values of resonant frequency of a parallel resonant circuit.

EXPERIMENT 2-1

MEASURING AC VOLTAGE

Alternating current is the most common form of electric current used in the United States. It is called *alternating* because it changes its direction periodically. The most frequently used unit of the time associated with ac is the second.

The number of ac cycles per second is known as *frequency*. Frequency is the number of times in 1 s that the ac moves from zero, reaches a peak in one direction, changes its direction, peaks in the opposite direction, and goes back to zero. These ac cycles per second (cps) are called *hertz* (Hz). The standard frequency of alternating current and voltage used in the United States is *60 Hz*. The period of this standard frequency is 0.0166 s.

Because alternating current, voltage, and power are constantly changing, two types of electric values are used in ac measurement. These are the *instantaneous value* and the *effective value*. Instantaneous values are used to describe the value of ac current, voltage, or power at any specified instant. The most common instantaneous value is *peak value:* the maximum value of voltage, current, or power during any cycle. Effective values are more well known because they are used to describe the amount of voltage, current, and power that can be counted on to produce light, heat, motion, or work of an electric nature. The effective values of ac produce the same amount of work as dc values. Effective values are also called root-mean-square (RMS) values.

Ohm's laws are used to compute ac values when the opposition to current is resistance only. Instantaneous values of voltage and current must be used to determine instantaneous power. Effective values must be used to determine effective power. The same procedure must be followed when determining voltage or current with a known resistance.

Instantaneous values of voltage, current, and power are converted to effective values, or vice versa, with the mathematical constants 0.707 or 1.41 in the following formulas:

Effective value = 0.707 × peak value

Peak value = 1.41 × effective value

Peak-to-peak values (the distance from peak to peak) of alternating current and voltage are computed with the following formula:

Peak-to-peak value = 2 × 1.41 × effective value

There is no peak-to-peak value for ac power.

OBJECTIVES

1. To study the characteristics of alternating current (ac).

2. To use a multimeter (VOM) to measure ac voltage.

EQUIPMENT

    Multimeter
    DC power supply or 6 V battery
    AC source: 30 V ac, center tapped with potentiometer adjust
    Single Pole Single Throw (SPST) switch
    Double Pole Double Throw (DPST) switch
    6 V lamp with socket
    Resistors: 100 Ω, 220 Ω, 300 Ω
    Connecting wires

PROCEDURE

1. Measure and record the resistance (R) of the filament of the 6 V lamp: $R =$ _____ Ω.

2. Construct the circuit in Fig. 2-1A.

3. Place the switch in position 1 and measure the dc voltage across the 6 V lamp. Record this voltage: _____ V dc.

4. Place the switch in position 2 and disconnect the VOM.

5. Connect a variable ac power supply to the circuit shown in Fig. 2-1B. (*Note:* Be sure to adjust the ac power supply to zero.) If you do not have a variable ac power supply, variable ac voltage may be obtained as shown from the circuit in Fig. 2-1C.

6. Prepare the VOM to measure ac voltage and connect it across the 6 V lamp.

7. Slowly adjust the variable ac voltage until the VOM reads the same ac voltage as the dc voltage recorded in step 3. Record this voltage: _____ V ac.

8. Slowly change the switch from position 2 to position 1 and back several times. How does this action affect the brightness of the bulb?
   _____
   _____

9. Using the data from steps 1 and 3, compute the following dc values when the switch is in position 1:

Current through the lamp: _____ A

Power dissipated by the lamp: _____ W

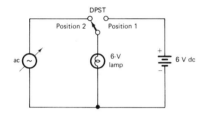

Fig. 2-1A. Simple lamp circuit.

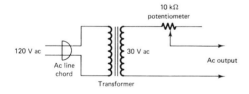

Fig. 2-1B. Lamp circuit with variable ac power supply.

Fig. 2-1C. Circuit to obtain variable ac voltage.

10. Using the data from steps 1 and 7, compute the following ac values when the DPST switch is in position 2:

Current through the lamp: _____ A

Power dissipated by the lamp: _____ W

11. How did the dc data in step 9 compare with the ac data in step 10? _____

_____

12. What conclusions can you reach when you compare ac effective values with identical dc values? _____

_____

13. Disconnect the circuit shown in step 5 and construct the circuit in Fig. 2-1D.

14. Using the VOM to measure ac voltage, complete Fig. 2-1E. Use the proper formulas for the computations.

ANALYSIS

1. Define the following terms:

    a. Hertz _____

    _____

    b. Frequency _____

    _____

    c. Period _____

    _____

    d. Cycle _____

    _____

    e. Effective ac values _____

    _____

    f. Instantaneous ac values _____

    _____

2. Draw a sine wave in the space to the right. Indicate its four quadrants, positive peak, negative peak, and time base (in degrees).

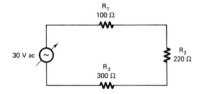

Fig. 2-1D. Series resistive circuit.

Resistor	Measured RMS voltage (V) across resistor	Computed peak voltage across resistor = V × 1.41	Computed p-p voltage across resistor = V × 2.82	Computed RMS power (P) across resistor = $\frac{V^2}{R}$	Computed peak power across resistor = P × 1.41
R_1					
R_2					
R_3					

Fig. 2-1E. AC values.

3. Convert the following effective values to peak and peak-to-peak values.

   6 V RMS = _____ V peak; _____ V p-p

   4 A RMS = _____ A peak; _____ A p-p

   10 V RMS = _____ V peak; _____ V p-p

   7 A RMS = _____ A peak; _____ A p-p

4. Convert the following peak values to effective values.

   12 W peak = _____ W RMS

   100 mA peak = _____ mA RMS

   2 mW peak = _____ mW RMS

5. How does the power produced by 10 V dc across 10 Ω compare with the power produced by 10 V ac RMS across the same resistance?
   _____
   _____

6. Why is a cycle said to consist of 360°?
   _____
   _____

7. Compute the periods for the following ac frequencies:

   1.2 kHz = _____ s

   10 kHz = _____ s

   1 MHz = _____ s

   0.6 MHz = _____ s

   60 Hz = _____ s

8. What is the standard frequency of the alternating current used in the United States? _____
   _____

9. What is alternating current? _____
   _____

10. If ac is continually changing its direction, why do the lights in your home not blink on and off?
    _____
    _____

Measuring AC 43

EXPERIMENT 2-2

MEASURING AC WITH AN OSCILLOSCOPE

An oscilloscope is an electronic instrument that allows various voltage waveforms to be analyzed visually. Somewhat like a television set, it produces an image on a screen when the controls are properly adjusted. The image, called the *trace,* is usually a line on a screen (CRT). A stream of electrons striking the phosphorous coating on the inside of the screen causes the coating.

An oscilloscope is used to measure frequency and voltage. It can also be used to determine current if used in conjunction with Ohm's law. The greatest application of an oscilloscope is allowing the operator to see the waveform or signals with which he or she is working.

OBJECTIVES

1. To become familiar with the controls of an oscilloscope.

2. To observe how the adjustments of these controls affect the waveform or trace displayed on the CRT.

3. To use an audio signal generator to supply variable ac voltage to a circuit.

EQUIPMENT

Oscilloscope
Audio signal generator
Resistor: 10 kΩ
Connecting wires
SPST switch
6 V battery

PROCEDURE

1. Turn on the oscilloscope and adjust the *intensity* and *focus* controls until a bright, sharp, straight-line trace appears on the screen. Use the *horizontal position* and *vertical position* controls to position the trace in the center of the screen. Adjust the *horizontal gain* until the line trace is long enough to extend from the extreme left to the extreme right side of the screen.

2. Connect the oscilloscope probe into the *vertical input* jack.

3. Construct the circuit in Fig. 2-2A. Adjust the signal generator to 60 Hz.

4. Connect the oscilloscope probes to points A and B in the circuit of Fig. 2-2A. (Be sure to connect the ground of the oscilloscope to the ground of the signal generator.)

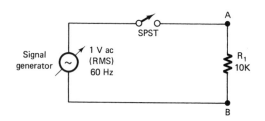

Fig. 2-2A. Simple test circuit.

5. Adjust the *vertical gain* controls until the height of the trace equals about 1 inch. Adjust the *horizontal stabilization* control until the trace becomes stable. Draw the waveform in the space below.

6. Slowly adjust the *vertical gain* controls both clockwise and counterclockwise. Describe how this action affects the trace.

   _____
   _____

7. Clockwise adjustment of *vertical attenuation* control caused the trace to _____
   _____.

   Counterclockwise adjustment of the *vertical attenuation* control caused the trace to _____
   _____.

8. Clockwise adjustment of the *vertical gain* control caused the trace to _____
   _____.

   Counterclockwise adjustment of the *vertical gain* control caused the trace to _____
   _____.

9. Adjust the *horizontal sweep selection* and *horizontal gain* controls both clockwise and counterclockwise. Describe how this action affects the trace. _____
   _____.

10. Clockwise adjustment of the *horizontal sweep selection* control caused the trace to _____
    _____.

    Counterclockwise adjustment of the *horizontal sweep selection* control caused the trace to
    _____.

11. Clockwise adjustment of the *horizontal gain* control caused the trace to _____
    _____.

    Counterclockwise adjustment of the *horizontal gain* control caused the trace to _____
    _____.

12. Adjust all of the oscilloscope controls to their maximum counterclockwise positions.

13. Open the SPST switch in the circuit illustrated in Fig. 2-2A. Adjust the signal generator to produce a signal of 1000 Hz.

14. Turn on the oscilloscope and close the switch.

15. Adjust the scope controls until a trace of two sine waves 2 cm in height is produced. Draw these waveforms here:

16. Adjust the scope controls until a trace of four sine waves 1 cm in height is produced. Draw these waveforms here:

17. Slowly adjust the frequency control of the signal generator both clockwise and counterclockwise. Describe what happens to the number of sine waves displayed by the oscilloscope. _____
   _____
   _____

18. Open the circuit's switch and disconnect the signal generator. In its place, connect a 6 V battery.

19. Change the controls of the scope to the dc vertical input. Adjust the *vertical attenuation* control and the *vertical gain* control to midrange. Position the trace on the screen as described in step 2.

20. With the scope's probes connected to points A and B in the circuit, close the SPST switch. Describe what happens to the line trace. Why does it happen? _____
   _____
   _____

ANALYSIS

1. What is the most important application of an oscilloscope? _____

2. Briefly describe the purpose of each of the following controls:
    a. Vertical gain _____
    b. Horizontal gain _____
    c. Intensity _____
    d. Focus _____
    e. Vertical position _____
    f. Horizontal position _____

3. What effect does adjusting the *horizontal sweep selection* switch have on the trace? _____

4. What does *synchronize* mean? _____

Unit 2 Examination
Measuring AC

Instructions: For each of the following questions, circle the answer that most correctly completes the statement.

1. The four values of an ac wave are peak, instantaneous, average, and effective. Which value is normally measured with an ac voltmeter?
   a. Peak
   b. Average
   c. Effective
   d. Instantaneous

2. When the volts/division switch is set at 5 V, a sine wave 3 divisions in height is seen on an oscilloscope. What is the peak-to-peak value of the input voltage?
   a. 3/5 V
   b. 3 V
   c. 5 V
   d. 15 V

3. A waveform is displayed on an oscilloscope. When the horizontal sweep speed is set at 10 µs/division, and one waveform spans a distance of 4 divisions, what is the frequency of the displayed voltage?
   a. 2.5 kHz
   b. 25 kHz
   c. 400 kHz
   d. 4 MHz

4. What horizontal sweep speed should be used to display one cycle of a 1 MHz signal on an oscilloscope?
   a. 0.05 µs/cm
   b. 0.1 µs/cm
   c. 0.5 µs/cm
   d. 1 µs/cm

5. If two complete cycles of 5 kHz sine waves are displayed on an oscilloscope, what is the horizontal sweep speed of the oscilloscope?
   a. 0.4 ms
   b. 2.5 ms
   c. 5 ms
   d. 10 ms

6. What is the time of one cycle of a 1 kHz waveform?
   a. 1×10^3 s
   b. 1 s
   c. 1×10^{-3} s
   d. 1×10^{-6} s

7. What is the frequency of the input signal if three complete cycles are observed spanning 6 divisions on an oscilloscope with a 10 ms horizontal sweep time?
   a. 200 Hz
   b. 500 Hz
   c. 2 kHz
   d. 50 kHz

8. What is the vertical height of a 6.3 V RMS signal displayed on an oscilloscope if the 2 V/cm position is selected for vertical gain?

   a. 3.15 cm   b. 4.41 cm
   c. 8.82 cm   d. 17.6 cm

9. What is the purpose of the calibration output on an oscilloscope?

   a. Calibrate the vertical (V/division)
   b. Calibrate the horizontal (time/division)
   c. Calibrate the rise time
   d. Calibrate the bandwidth

10. When measuring ac voltage with a VOM

    a. Polarity must be observed.
    b. The meter must be recalibrated before each measurement.
    c. Polarity is not observed.
    d. The "dc volts" setting is used to measure RMS values.

True or false: Place either T or F in each blank.

_____ 11. An oscilloscope is an instrument that allows its operator to analyze voltage waveforms visually.

_____ 12. The image or waveform displayed by an oscilloscope is sometimes called the trace.

_____ 13. An oscilloscope displays the voltage waveforms on three axes: vertical, horizontal, and time.

_____ 14. The brightness of a waveform displayed by an oscilloscope is controlled with the focus control.

_____ 15. The horizontal gain control of an oscilloscope is used to control the display height of the waveform.

_____ 16. The intensity control is used to adjust the displayed waveform to cause it to become clear and sharp.

_____ 17. The vertical position control is used to adjust the displayed waveform up and down.

_____ 18. The inside of the CRT is coated with potassium hydroxide and glows when struck with the beam of electrons.

_____ 19. The oscilloscope can be used to measure frequency as well as voltage.

_____ 20. The horizontal gain control is used to adjust the waveform to the left or right.

UNIT 3

Resistance, Inductance, and Capacitance in AC Circuits

Alternating current electronic circuits are similar in many ways to direct current (dc) circuits. AC circuits are classified by their electric characteristics (resistive, inductive, or capacitive). All electronic ac circuits are resistive, inductive, capacitive, or a combination of these characteristics. The operation of each type of electronic circuit is different. The nature of ac causes certain circuit properties to exist.

> *UNIT OBJECTIVES*
>
> Upon completion of this unit, you should be able to:
>
> 1. Define and calculate impedance.
> 2. Draw diagrams that illustrate the phase relation between current and voltage in a capacitive circuit.
> 3. Define capacitive reactance.
> 4. Solve problems using the capacitive reactance formula.
> 5. Define impedance.
> 6. Calculate impedance of series and parallel resistive/capacitive circuits.
> 7. Determine current in *RC* circuits.
> 8. Solve phase-angle problems using a calculator or a table of trigonometric functions.
> 9. Explain the relation between ac voltages and current in a series resistive circuit.
> 10. Understand the effect of capacitors in series and parallel.
> 11. Explain the characteristics of a series *RC* circuit.
> 12. Solve Ohm's law problems for ac circuits.
> 13. Understand and solve problems involving true power, apparent power, power factor, and reactive power.

14. Solve problems involving inductive reactance, impedance, resistance, phase angle, apparent power, true power, power factor, and reactive power.
15. Solve for values of impedance, current, voltage, power, and phase angle in series and parallel *RLC* circuits.
16. Draw diagrams illustrating phase relations between voltage and current in an inductive circuit.
17. Define and calculate inductive reactance.
18. Solve problems involving inductive reactance, frequency, and inductance.

Important Terms

The following terms provide a review of resistance, inductance, and capacitance in ac circuits:

Admittance (Y) The total ability of an ac circuit to conduct current, measured in siemens (S) or ohms (Ω); the inverse of impedance: $Y = 1/Z$.

Angle of lead or lag The phase angle between applied voltage and total current flow in an ac circuit (in degrees). In an inductive circuit, voltage leads current; in a capacitive circuit, voltage lags current.

Apparent power (volt-amperes) The power delivered to an ac circuit; applied voltage times total current in an ac circuit.

Capacitance (C) The property of a circuit or device to oppose changes in voltage due to energy stored in an electrostatic field.

Capacitive reactance (X_C) The opposition to the flow of ac current caused by a capacitive device (measured in ohms [Ω]).

Capacitor A device that has capacitance and usually is made of two metal plates separated by a dielectric material (insulator).

Conductance (G) The ability of the resistance of an ac circuit to conduct current, measured in siemens (S) or mhos; the inverse of resistance: $G = 1/R$.

Dielectric The insulating material placed between the metal plates of a capacitor.

Dielectric constant A number that represents the ability of a dielectric material to develop an electrostatic field compared with air, which has a value of 1.0.

Electrolytic capacitor A capacitor that has a positive plate made of aluminum and a dry paste or liquid used to form the negative plate.

Electrostatic field The field developed around a material because of the energy of an electric charge, as is exhibited by capacitors.

Farad (F) The unit of measurement of capacitance; produced when a charge of 1 coulomb (C) causes a potential of 1 V to be developed across two points.

Henry (H) The unit of measurement of inductance; produced when a voltage of 1 V is induced as the current through a coil is changing at a rate of one ampere per second.

Impedance (Z) The total opposition to current flow in an ac circuit, a combination of resistance (R) and reactance (X) in a circuit (measured in ohms [Ω]):

$$Z = \sqrt{R^2 + X^2}$$

Inductance (L) The property of a circuit to oppose changes in current due to energy stored in a magnetic field.

Inductive circuit A circuit that has one or more inductors or has the property of inductance, such as an electric motor circuit.

Inductive reactance (X_L) The opposition to current flow in an ac circuit caused by an inductance (L); measured in ohms [Ω]: $X_L = 2\pi \times f \times L$

Inductors Coils of wire or windings that possess the property of inductance and are used in a circuit to cause inductance to be present.

Lagging phase angle The angle by which current *lags* voltage (or voltage *leads* current) in an inductive circuit.

Leading phase angle The angle by which current *leads* voltage (or voltage *lags* current) in a capacitive circuit.

Mho *Ohm* spelled backward; a unit of measurement for conductance, susceptance, and admittance in ac circuits that is being replaced by the unit *siemen* (S).

Mica capacitor A capacitor made of metal foil plates separated by a mica dielectric.

Mutual inductance (M) The situation in which two coils are located close together so that the magnetic fluxes of the coils affect one another in terms of total inductance properties.

Power factor (pf) The ratio of power converted (true power) in an ac circuit and the power delivered to the circuit.

$$\text{(Apparent power) pf} = \frac{\text{true power (watts)}}{\text{apparent power (volt-amperes)}}$$

Radian The measure of an angle formed by rotating the radius of a circle until the arc made by the end of the radius is the same length as the radius; 1 radian (rad) = 57.3°; the circumference of a circle is 2π (6.28) radians.

Reactance (X) The opposition to current flow in an ac circuit caused by inductance (inductive reactance, X_L) or capacitance (capacitive reactance, X_C).

Reactive circuit An ac circuit that has the property of inductance or capacitance or both.

Reactive power (VAR) The "unused" power of an ac circuit that has inductance or capacitance; the power absorbed by the magnetic or electrostatic field of a reactive circuit.

Resistance The opposition to current flow in a circuit caused by a resistive device.

Resistive circuit A circuit in which the only opposition to current flow is resistance; a nonreactive circuit.

Siemen (S) See *mho*.

Susceptance (B) The ability of inductance (B_L) or capacitance (B_C) to pass ac current; measured in siemens or mhos; $B_L = 1/X_L$ and $B_C = 1/X_C$.

True power (watts) The power actually converted to another form of energy in an ac circuit.

VAR (volt-amperes-reactive) The unit of measurement of reactive power.

Vector A straight line that indicates a quantity with magnitude and direction.

Volt-ampere (VA) The unit of measurement of apparent power.

Working voltage A rating of capacitors; the maximum voltage that can be placed across the plates of a capacitor without damage.

Resistive AC Circuits

The simplest type of ac circuit is a resistive circuit, such as the one shown in Fig. 3-1. The purely resistive circuit offers the same type of opposition to ac as it does to pure dc sources. In dc circuits, the following relations exist:

$$\text{Voltage } (V) = \text{current } (I) \times \text{resistance } (R)$$

$$\text{Current } (I) = \frac{\text{voltage } (V)}{\text{resistance } (R)}$$

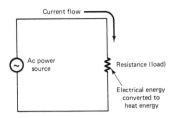

Fig. 3-1. Resistive ac circuit.

$$\text{Resistance } (R) = \frac{\text{voltage } (V)}{\text{current } (I)}$$

$$\text{Power } (P) = \text{voltage } (V) \times \text{current } (I)$$

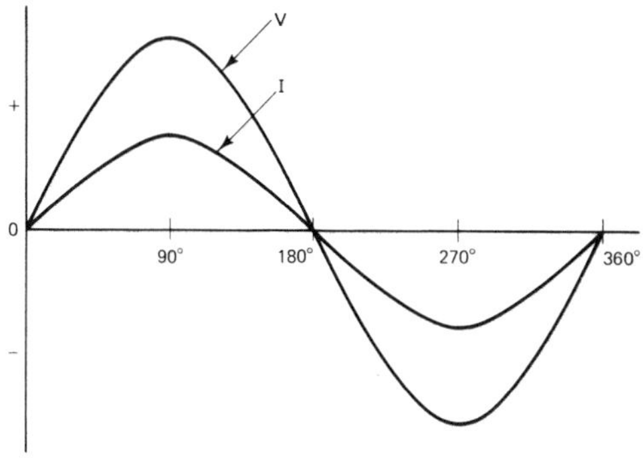

These basic electric relations show that when voltage is increased, the current in the circuit increases proportionally. As resistance is increased, the current in the circuit decreases. The waveforms in Fig. 3-2 show that the voltage and current in a purely resistive circuit with ac applied are in phase. An in-phase relation exists when the minimum and maximum values of both voltage and current occur at the same time interval. The power converted by the circuit is a product of voltage times current ($P = V \times I$). The power curve is shown in Fig. 3-3. The behavior of an ac circuit that contains only resistance is very similar to that of a dc circuit. Purely resistive circuits are seldom encountered in the design of electric power systems, although some devices are primarily resistive in nature.

Fig. 3-2. Voltage and current waveforms of a resistive ac circuit.

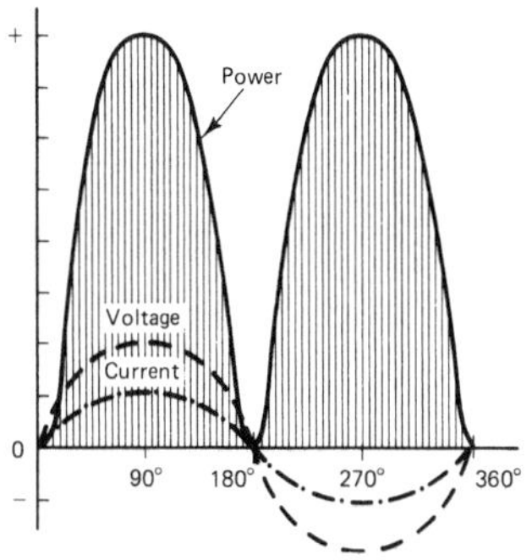

Fig. 3-3. Power curve for a resistive ac circuit.

Inductive Circuits

The property of inductance (L) is commonly encountered in electronic circuits. This circuit property, shown in Fig. 3-4, adds more complexity to the relation between voltage and current in an ac circuit. All motors, generators, and transformers exhibit the property of inductance. This property is evident because of a *counter-electromotive force (CEMF)*, which is produced when a magnetic field is developed around a coil of wire. The magnetic flux produced around the coils affects circuit action. Thus the inductive property (CEMF) produced by a magnetic field offers opposition to change in the current flow in a circuit.

The opposition to change of current is evident in Fig. 3-4b. In an inductive circuit, *voltage leads current* or *current lags voltage*. In a purely inductive circuit (contains no resistance), voltage leads the current by 90° (Fig. 3-5), and no power is converted in the circuit. However, because all actual circuits have resistance, the inductive characteristic of a circuit might typically cause the condition shown in Fig. 3-6. The voltage is leading the current by 30°. The angular separation between voltage and current is called the *phase angle*. The phase angle increases as the inductance of the circuit increases. This type of circuit is called a *resistive-inductive (RL) circuit*.

In terms of power conversion, a purely inductive circuit does not convert any power in a circuit. All ac power is delivered back to the power source. Refer to Fig. 3-5 and look at points *A* and *B* on the waveforms. These points show that at the peak of each waveform, the corresponding value of the other waveform is zero. The power curves shown are equal and opposite in value and cancel each other out. When both voltage and current are positive, the power is positive because the product of two positive

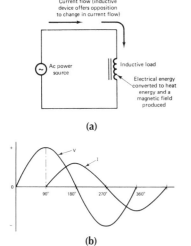

Fig. 3-4. Voltage and current waveforms of a purely inductive ac circuit. (a) Circuit. (b) Waveforms.

values is positive. When one value is positive and the other is negative, the product of the two values is negative; therefore the power converted is negative. Negative power means that electric energy is being returned from the load device to the power source without being converted to another form of energy. Therefore the power converted in a purely inductive circuit (90° phase angle) equals zero.

Compare the purely inductive waveforms of Fig. 3-5 to those of Fig. 3-6. In the practical RL circuit, part of the power supplied from the source is converted in the circuit. Only during the intervals from 0° to 30° and from 180° to 210° does negative power result. The remainder of the cycle produces positive power; therefore, most of the electric energy supplied by the source is converted to another form of energy.

Any inductive circuit exhibits the property of inductance (L), which is the opposition to a change in current flow in a circuit. This property is found in coils of wire (*inductors*) and in rotating machinery and transformer windings. Inductance also is present to some extent in electric power transmission and distribution lines. The unit of measurement for inductance is the henry (H). A circuit has 1 H of inductance when a current change of 1 ampere per second (A/s) produces an induced CEMF of 1 V.

In an inductive circuit with ac applied, an opposition to current flow is produced by the inductance. This type of opposition is known as *inductive reactance* (X_L). The inductive reactance of an ac circuit depends on the inductance (L) of the circuit and the rate of change of current. The frequency of the applied ac establishes the rate of change of the current. Inductive reactance (X_L) may be expressed as follows:

$$X_L = 2\pi f L$$

where

X_L = the inductive reactance in ohms (Ω)

2π = 6.28, the mathematical expression for one sine wave of ac (0°–360°)

f = the frequency of the ac source in hertz (Hz)

L = the inductance of the circuit in henrys

Mutual Inductance

When inductors are connected together, a property called *mutual inductance (M)* must be considered. Mutual inductance is the magnetic field interaction or flux linkage between coils. The amount of flux linkage is called the *coefficient of coupling (k)*. If all the lines of force of one coil cut across a nearby coil, the condition is called *unity coupling*. There are many possibilities, and they are determined by coil placement of coupling between

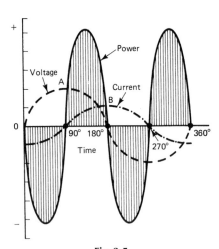

Fig. 3-5.
Power curve for a purely inductive circuit.

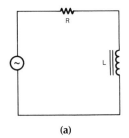

(a)

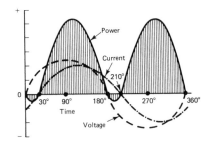

(b)

Fig. 3-6. Resistive-inductive (*RL*) circuit and its waveforms. (a) *RL* ac circuit. (b) Voltage and current waveforms.

coils. The amount of mutual inductance between coils is found with the following formula:

$$(M) = k \times \sqrt{L_1 \times L_2}$$

The term k is the coefficient of coupling, which gives the amount of coupling. L_1 and L_2 are the inductance values of the coils. Mutual inductance should be considered when two or more coils are connected together.

Inductors in Series and Parallel

Inductors may be connected in series or parallel. When inductors are connected to prevent the magnetic field of one from affecting the others, these formulas are used to find *total inductance* (L_T).

1. Series inductance

$$L_T = L_1 + L_2 + L_3 + \ldots L_n$$

2. Parallel inductance

$$\frac{1}{L_T} = \frac{1}{L_1} + \frac{1}{L_2} + \frac{1}{L_3} + \cdots \frac{1}{L_n}$$

where $L_1, L_2, L_3, \ldots$ are inductance values measured in henrys.

When inductors are connected so that the magnetic field of one affects the other, mutual inductance increases or decreases total inductance. The effect of mutual inductance depends on the physical positioning of the inductors. The distance apart and the direction in which they are wound affect mutual inductance. Inductors are connected in series or parallel with an aiding or opposing mutual inductance (M). The following formulas are used to find total inductance (L_T):

1. Series aiding

$$L_T = L + L_2 + 2M$$

2. Series opposing

$$L_T = L_1 + L_2 - 2M$$

3. Parallel aiding

$$\frac{1}{L_T} = \frac{1}{L_1 + M} + \frac{1}{L_2 + M}$$

4. Parallel opposing

$$\frac{1}{L_T} = \frac{1}{L_1 - M} + \frac{1}{L_2 - M}$$

L_1 and L_2 are the inductance values, and M is the value of mutual inductance.

Capacitive Circuits

Figure 3-7 shows a capacitive device connected to an ac source. Whenever two conductive materials (plates) are separated by an insulating (dielectric) material, the property of capacitance is exhibited. Capacitors can store an electric charge. They also have many applications in electric power systems.

The operation of a capacitor in a circuit depends on its ability to charge and discharge. When a capacitor charges, an excess of electrons (negative charge) accumulates on one plate and a deficiency of electrons (positive charge) occurs on the other plate. Capacitance (C) is determined by the area of the conductive material, the thickness of the dielectric, and the type of insulating material. Capacitance is directly proportional to the plate area and inversely proportional to the distance between the plates. These factors are expressed with the following formula:

$$C = \frac{0.0885ka}{t}$$

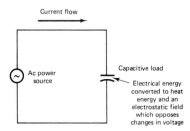

Fig. 3-7. Capacitive ac circuit.

where C is capacitance in picofarads (pF), k is the dielectric constant, a is the area of one plate in square centimeters (cm^2), and t is the dielectric thickness in centimeters (cm).

Example

Compute the capacitance of a capacitor that has two metal foil plates that are 2.5 m wide and 250 cm long. The dielectric material is wax paper. Wax paper has a dielectric constant of 2. The thickness of the wax paper is 0.025 cm.

Substituting these values in the capacitance formula gives the following:

$$C = 0.0885 \times 2 \times 625 \text{ cm}/0.025 \text{ cm}$$

$$= 4425 \text{ pF or } 0.004425 \text{ μF}$$

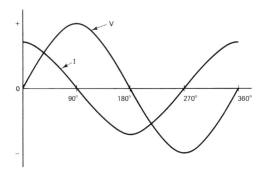

Fig. 3-8. Voltage and current waveforms of a purely capacitive ac circuit.

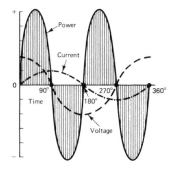

Fig. 3-9. Power curves for a purely capacitive ac circuit.

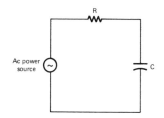

Fig. 3-10. Resistive-capacitive (RC) circuit.

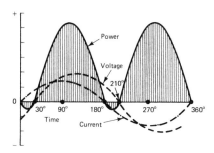

Fig. 3-11. Waveforms of an RC circuit.

The fundamental unit of capacitance is the farad (F). A capacitance of 1 F results when a potential of 1 V causes an electric charge of 1 coulomb (C) to accumulate on a capacitor. Because a farad is a very large unit, measurements in microfarads (μF) and picofarads (pF) are ordinarily assigned to capacitors.

If a dc is applied to a capacitor, the capacitor charges to the value of that dc voltage. After the capacitor is fully charged, it blocks the flow of dc. However, if ac is applied to a capacitor, the changing value of current causes the capacitor alternately to charge and discharge. In a purely capacitive circuit, the situation shown in Fig. 3-8 would exist. The greatest amount of current would flow in a capacitive circuit when the voltage changed most rapidly. The most rapid change in voltage occurs at the 0° and 180° positions, where polarity changes. At these positions, maximum current is developed in the circuit. When the rate of change of the voltage value is slow, such as near the 90° and 270° positions, a small amount of current flows. Figure 3-8 shows that current leads voltage by 90° in a purely capacitive circuit and that voltage lags current by 90°. Because a 90° phase angle exists, no power would be converted in this circuit, just as no power was developed in the purely inductive circuit. As shown in Fig. 3-9, the positive and negative power waveforms cancel one another.

Because all circuits contain some resistance, a more practical circuit is the resistive-capacitive (RC) circuit shown in Fig. 3-10. In an RC circuit, current leads voltage by a phase angle between 0° and 90°. As capacitance increases with no corresponding increase in resistance, the phase angle becomes greater. The waveforms of Fig. 3-11 show an RC circuit in which current leads voltage by 30°. This circuit is similar to the RL circuit in Fig. 3-6. Power is converted in the circuit except during the 0° to 30° interval and the 180° to 210° interval. In the RC circuit shown, most of the electric energy supplied by the source is converted to another form of energy in the load.

Because of the electrostatic field that develops around a capacitor, an opposition to the flow of ac exists. This opposition is known as *capacitive reactance* (X_C). Capacitive reactance is expressed as follows:

$$X_C = \frac{1}{2\pi f C}$$

where

X_C = capacitive reactance in ohms

2π = the mathematical expression of one sine wave of ac (0 - 360°)

f = the frequency of the source in hertz

C = the capacitance in farads

Capacitors in Series

Adding capacitors in series has the same effect as increasing the distance between plates. This reduces total capacitance. Total capacitance (C_T) is found in the same way as parallel resistance. The reciprocal formula is used, as follows:

$$\frac{1}{C_T} + \frac{1}{C_1} + \frac{1}{C_2} + \frac{1}{C_3} + \cdots \frac{1}{C_n}$$

Capacitors in Parallel

Capacitors in parallel resemble one capacitor with its plate area increased. Doubling plate area doubles capacitance. Capacitance in parallel is found by adding individual values, just as with series resistors:

$$C_T = C_1 + C_2 + C_3 + \ldots C_n$$

Vector (Phasor) Diagrams

Figure 3-12 shows a vector diagram for each circuit condition. Vectors are straight lines that have a specific direction (angle with respect to a reference direction) and length (magnitude). They may be used to represent voltage or current values. An understanding of *vector diagrams* (sometimes called *phasor diagrams*) is important when dealing with ac. Rather than using waveforms to show phase relations, it is possible to use a vector or phasor representation.

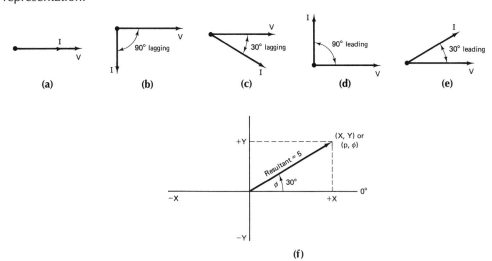

Fig. 3-12. Vector diagram showing voltage and current relationship of ac circuits: (a) resistive (R) circuit; (b) purely inductive (L) circuit; (c) RL circuit; (d) purely capacitive (C) circuit; (e) RC circuit; (f) rectangular and polar coordinates.

Ordinarily when beginning a vector diagram one draws a horizontal line with its left end as the reference point. Rotation in a counterclockwise direction from the reference point is considered a positive direction. In the diagrams in Fig. 3-12, the voltage vector is the reference. For inductive circuits, the current vector is drawn in a clockwise direction, indicating a lagging condition. A leading condition is shown for the capacitive circuits through use of a current vector drawn in a counterclockwise direction from the voltage vector.

Vectors and Coordinate Systems

Many ac circuit problems require the use of vectors and coordinate systems. An arrowhead is used on the end of the vector to indicate the direction toward which a force is acting. Most vector analysis deals with determining the magnitude and direction of a *resultant* vector produced by two or more similar vectors acting on a point (see Fig. 3-12).

Rectangular Coordinate System

A rectangular coordinate system is useful when the rectangular components of a vector are needed. The rectangular components can be read directly by means of observing or measuring the horizontal component, X, and the vertical component, Y. The coordinates of the resultant vector in Fig. 3-12f are (X, Y).

Polar Coordinate System

A polar coordinate system is most useful when the magnitude and direction of the vector are required. The magnitude ρ (Greek letter rho) is given directly by the length of the vector. The direction is given as the angle θ (Greek letter theta). The resultant vector in Fig. 3-12f is represented as $\rho \angle \theta$ or $5 \angle 30°$. The $\angle$ symbol represents an angle. The example is read as 5 at an angle of 30°.

Series AC Circuits

In any series ac circuit, the current is the same in all parts of the circuit. The voltages must be added with a *voltage triangle*. The impedance (Z) of a series ac circuit is found with an *impedance triangle*. Power values are found with a *power triangle*.

Series *RL* Circuits

Series *RL* circuits often are present in electronic equipment. When an ac voltage is applied to a series *RL* circuit, the current is the same through each part. The voltage drops across each component are distributed according to the values of resistance (R) and inductive reactance (X_L) in the circuit.

The total opposition to current flow in any ac circuit is called *impedance* (Z). Both resistance and reactance in an ac circuit oppose current flow. The impedance of a series *RL* circuit is found with either of the following formulas:

$$Z = \frac{V}{I}$$

or

$$Z = \sqrt{R^2 + X_L^2}$$

The impedance of a series *RL* circuit is found with the *impedance triangle*. This right triangle is formed by the three quantities that oppose ac. A triangle also is used to compare voltage drops in series *RL* circuits. Inductive voltage (V_L) leads resistive voltage (V_R) by 90°. V_A is the voltage applied to the circuit. Because these values form a right triangle, the value of V_A may be found with the following formula:

$$V_A = \sqrt{V_R^2 + V_L^2}$$

An example of a series *RL* circuit is shown in Fig. 3-13. Appendix B should be reviewed, if necessary, to gain a better understanding of trigonometry for ac circuit problems.

Series *RC* Circuits

Series *RC* circuits have many uses. This type of circuit is similar to a series *RL* circuit. In a capacitive circuit, current leads voltage. The reactive values of *RC* circuits act in the opposite directions to *RL* circuits. An example of a series *RC* circuit is shown in Fig. 3-14.

Series *RLC* Circuits

Series *RLC* circuits have resistance (R), inductance (L), and capacitance (C). The total reactance (X_T) is found by means of subtracting the smaller reactance (X_L or X_C) from the larger one. Reactive voltage (V_x) is found by means of obtaining the difference between V_L and V_C. The effects of capacitance and inductance are 180° out of phase with each other. Right triangles are used to show the simplified relations of the circuit values. An example of a series *RLC* circuit is shown in Fig. 3-15.

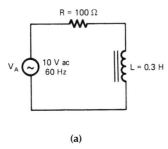

(a)

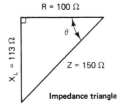

Impedance triangle

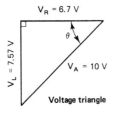

Voltage triangle

(c)

Fig. 3-13. Example of a series *RL* circuit. (a) Circuit. (b) Procedure for finding circuit values. (c) Circuit triangles.

(b) Finding circuit values:

1. Find inductive reactance (X_L):
$$X_L = 2\pi \cdot f \cdot L$$
$$= 6.28 \times 60 \times 0.3$$
$$= 113 \, \Omega$$

2. Find impedance (Z):
$$Z = \sqrt{R^2 + X_L^2}$$
$$= \sqrt{100^2 + 113^2}$$
$$= \sqrt{10{,}000 + 12{,}769}$$
$$= \sqrt{22{,}769}$$
$$= 150 \, \Omega$$

3. Check to be sure that Z is larger than R or X_L.

4. Find total current (I_T):
$$I_T = \frac{V}{Z} = \frac{10 \, V}{150 \, \Omega} = 0.067 \, A$$

5. Find the voltage across R (V_R):
$$V_R = I \times R$$
$$= 0.067 \, A \times 100 \, \Omega$$
$$= 6.7 \, V$$

6. Find the voltage across L (V_L):
$$V_L = I \cdot X_L$$
$$= 0.067 \, A \times 133 \, \Omega$$
$$= 7.57 \, V$$

7. Check to see that
$$V_A = \sqrt{V_R^2 + V_L^2}$$
$$10 \, V = \sqrt{6.7 \, V^2 + 7.57 \, V^2}$$
$$= \sqrt{44.89 + 57.3}$$
$$= \sqrt{102.2}$$
$$\approx 10.1 \, V^*$$

*(approximately equal to)

8. Find circuit phase angle (θ):
$$\text{cosine } \theta = \frac{\text{adjacent}}{\text{hypotenuse}} = \frac{V_R}{V_A} = \frac{6.7}{10}$$
$$= 0.67$$
$$= \text{inverse cosine } 0.67$$
$$= 48°$$

PARALLEL AC CIRCUITS

Parallel ac circuits have many applications. The basic formulas used with parallel ac circuits are different from those used with series circuits. The impedance (Z) of a parallel circuit is less than individual branch values of resistance, inductive reactance, or capacitive reactance. There is no impedance triangle for parallel

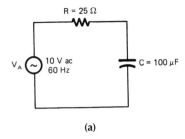

(a)

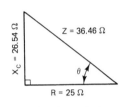

Impedance triangle

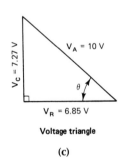

Voltage triangle

(c)

Fig. 3-14. Example of a series RC circuit. (a) Circuit. (b) Procedures for finding circuit values. (c) Circuit triangles.

(b) Finding circuit values:

1. Find capacitive reactance (X_C):

$$X_C = \frac{1}{2\pi \cdot f \cdot C} \text{ or } \frac{1,000,000}{2\pi \cdot f \cdot C}$$
$$\text{(in farads)} \quad \text{(in } \mu F\text{)}$$

$$= \frac{1,000,000}{6.28 \times 60 \times 100}$$

$$= 26.54 \, \Omega$$

2. Find impedance (Z):

$$Z = \sqrt{R^2 + X_C^2}$$
$$= \sqrt{25^2 + 26.54^2}$$
$$= \sqrt{625 + 704.37}$$
$$= \sqrt{1329.37}$$
$$= 36.46 \, \Omega$$

3. Check to be sure that Z is larger than R or X_C.

4. Find total current (I_T):

$$I_T = \frac{V}{Z} = \frac{10 \text{ V}}{36.46 \, \Omega} = 0.274 \text{ A}$$

5. Find the voltage across R (V_R):

$$V_R = I \times R$$
$$= 0.274 \text{ A} \times 25 \, \Omega$$
$$= 6.85 \text{ V}$$

6. Find the voltage across C (V_C):

$$V_C = I \cdot X_C$$
$$= 0.274 \text{ A} \times 26.54 \, \Omega$$
$$= 7.27 \text{ V}$$

7. Check to see that
$$V_A = \sqrt{V_R^2 + V_C^2}:$$
$$10 \text{ V} = \sqrt{6.85 \text{ V}^2 + 7.27 \text{ V}^2}$$
$$= \sqrt{46.92 + 52.88}$$
$$= \sqrt{99.8}$$
$$\approx 9.99 \text{ V}$$

8. Find the circuit phase angle (θ):

$$*\sin \theta = \frac{\text{opposite}}{\text{hypotenuse}} = \frac{X_C}{Z} = \frac{26.54}{36.46}$$
$$= 0.727$$
$$\theta = \text{inverse sine } 0.727$$
$$= 47°$$

*Any trig function can be used with either triangle.

circuits because Z is smaller than R, X_L, or X_C. A right triangle is drawn to show the currents in the branches of a parallel circuit.

The voltage of a parallel ac circuit is the same across each branch. The currents through the branches of a parallel ac circuit are shown with a right triangle called a *current triangle*. The current through the capacitor (I_C) is shown leading the current through the resistor (I_R) by 90°. The current through the inductor

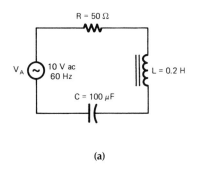

(a)

(b) Finding circuit values:

1. Find inductive reactance (X_L):

   $X_L = 2\pi \cdot f \cdot L$

   $= 6.28 \times 60 \times 0.2$

   $= 75.36 \ \Omega$

2. Find capacitive reactance (X_C):

   $X_C = \dfrac{1{,}000{,}000}{2\pi \cdot f \cdot C \ (\mu F)}$

   $= \dfrac{1{,}000{,}000}{6.28 \times 60 \times 100}$

   $= 26.54 \ \Omega$

3. Find total reactance (X_T):

   $X_T = X_L - X_C$

   $= 75.36 \ \Omega - 26.54 \ \Omega$

   $= 48.82 \ \Omega$

4. Find impedance (Z):

   $Z = \sqrt{R^2 + X_T^2}$

   $= \sqrt{50^2 + 48.82^2}$

   $= \sqrt{2500 + 2383.4}$

   $= \sqrt{4{,}883.4}$

   $= 69.88 \ \Omega$

5. Check to be sure that Z is larger than X_T or R.

6. Find total current (I_T):

   $I_T = \dfrac{V}{Z} = \dfrac{10 \ V}{69.88 \ \Omega} = 0.143 \ A$

7. Find voltage across R (V_R):

   $V_R = I \times R$

   $= 0.143 \ A \times 50 \ \Omega$

   $= 7.15 \ V$

8. Find voltage across L (V_L):

   $V_L = I \cdot X_L$

   $= 0.143 \ A \times 75.36 \ \Omega$

   $= 10.78 \ V$

9. Find voltage across C (V_C):

   $V_C = I \cdot X_C$

   $= 0.143 \ A \times 26.54 \ \Omega$

   $= 3.8 \ V$

10. Find total reactive voltage (V_X):

    $V_X = V_L - V_C$

    $= 10.78 \ V - 3.8 \ V$

    $= 6.98 \ V$

11. Check to see that

    $V_A = \sqrt{V_R^2 + V_X^2}$

    $= \sqrt{7.15 \ V^2 + 6.98 \ V^2}$

    $10 \ V = \sqrt{51.12 + 48.72}$

    $= \sqrt{99.84}$

    $\approx 9.99 \ V$

12. Find circuit phase angle (θ):

    $\tan \theta = \dfrac{\text{opposite}}{\text{adjacent}} = \dfrac{X_T}{R} = \dfrac{48.82}{50}$

    $= 0.976$

    $= \text{inverse tangent } 0.976$

    $= 44°$

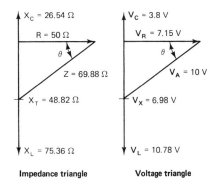

Impedance triangle   Voltage triangle

(c)

Fig. 3-15. Example of a series *RL* circuit: (a) Circuit.
(b) Procedure for finding circuit values.
(c) Circuit triangles.

(I_L) is shown lagging I_R by 90°. I_L and I_C are 180° out of phase. They are subtracted to find the total reactive current (I_X). Because these values form a right triangle, the total current may be found with the following formula:

$$I_T = \sqrt{I_R^2 + I_X^2}$$

This method is used to find currents in parallel *RL, RC,* or *RLC* circuits.

When components are connected in parallel, finding impedance is more difficult. An impedance triangle cannot be used. A method that can be used to find impedance is to use an *admittance triangle*. The following quantities are plotted on the triangle: admittance: $Y = 1/Z$; conductance: $G = 1/R$; inductive susceptance: $B_L = 1/X_L$; and capacitive susceptance: $B_C = 1/X_C$. These quantities are the inverse of each type of opposition to ac. Because total impedance (Z) is the smallest quantity in a parallel ac circuit, its reciprocal ($1/Z$) becomes the largest quantity on the admittance triangle (just as ½ is larger than ¼). The values are in siemens (S) or mho (ohm spelled backward).

Parallel ac circuits are similar in several ways to series ac circuits. Study the examples of Figs. 3-16 through 3-18.

Power in AC Circuits

Power values in ac circuits are found with *power triangles*, as shown in Fig. 3-19. The power delivered to an ac circuit is called *apparent power*. It equals voltage times current. The unit of measurement is the volt-ampere (VA) or kilovolt-ampere (kVA). Meters are used to measure apparent power in ac circuits. Apparent power is the voltage applied to a circuit multiplied by the total current flow. The power converted to another form of energy by the load is measured with a wattmeter. This actual power converted is called *true power*. The ratio of true power converted in a circuit to apparent power delivered to an ac circuit is called the *power factor* (pf). The power factor is found with the following formula:

$$\text{pf} = \frac{\text{true power (watts)}}{\text{apparent power (volt-amperes)}}$$

or

$$\% \text{ pf} = \frac{W}{VA} \times 100$$

W is the true power in watts and *VA* is the apparent power in volt-amperes. The highest power factor of an ac circuit is 1.0, or 100%. This value is called the *unity power factor*.

The phase angle (θ) of an ac circuit determines the power factor. The phase angle is the angular separation between voltage

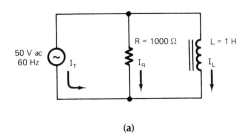

(a)

(b) Finding circuit values:

1. Find inductive reactance (X_L):

   $X_L = 2\pi \cdot f \cdot L$

   $= 6.28 \times 60 \times 1$

   $= 376.8\ \Omega$

2. Find current through R (I_R):

   $I_R = \dfrac{V}{R} = \dfrac{50\ V}{1000\ \Omega} = 0.05\ A$

3. Find current through L (I_L):

   $I_L = \dfrac{V}{X_L} = \dfrac{50\ V}{376.8\ \Omega} = 0.133\ A$

4. Find total current (I_T):

   $I_T = \sqrt{I_R^2 + I_L^2}$

   $= \sqrt{0.05^2 + 0.133^2}$

   $= \sqrt{0.0025 + 0.0177}$

   $= \sqrt{0.0202}$

   $= 0.142\ A$

5. Check to see that I_T is larger than I_R or I_L.

6. Find impedance (Z):

   $Z = \dfrac{V}{I_T} = \dfrac{50\ V}{0.142\ A} = 352.1\ \Omega$

7. Check to see that Z is less than R or X_L.

8. Find circuit phase angle (θ):

   $\sin \theta = \dfrac{\text{opposite}}{\text{hypotenuse}} = \dfrac{I_L}{I_T} = \dfrac{0.133}{0.142}$

   $= 0.937$

   $\theta = \text{inverse sine}\ 0.937$

   $= 70°$

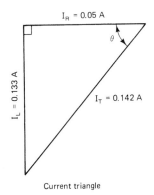

Current triangle

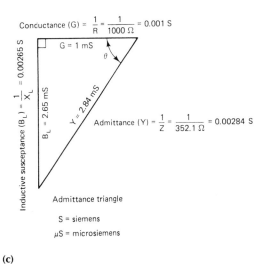

Admittance triangle

S = siemens
μS = microsiemens

(c)

Fig. 3-16. Example of a parallel *RL* circuit. (a) Circuit. (b) Procedure for finding circuit values. (c) Circuit triangles.

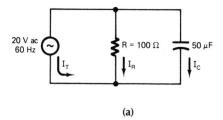

(a)

(b) Finding circuit values:

1. Find capacitive reactance (X_C):

$$X_C = \frac{1{,}000{,}000}{2\pi \cdot f \cdot C\,(\mu F)}$$

$$= \frac{1{,}000{,}000}{6.28 \times 60 \times 50}$$

$$= 53\ \Omega$$

2. Find current through R (I_R):

$$I_R = \frac{V}{R} = \frac{20\ V}{100\ \Omega} = 0.2\ A$$

3. Find current through C (I_C):

$$I_C = \frac{V}{X_C} = \frac{20\ V}{53\ \Omega} = 0.377\ A$$

4. Find total current (I_T):

$$I_T = \sqrt{I_R^2 + I_C^2}$$
$$= \sqrt{0.2^2 + 0.377^2}$$
$$= \sqrt{0.04 + 0.142}$$
$$= \sqrt{0.182}$$
$$= 0.427\ A$$

5. Check to see that I_T is larger than I_R or I_C.

6. Find impedance (Z):

$$Z = \frac{V}{I_T} = \frac{20\ V}{0.427\ A} = 46.84\ \Omega$$

7. Check to see that Z is less than R or X_C.

8. Find circuit phase angle (θ):

$$\cosine\ \theta = \frac{\text{adjacent}}{\text{hypotenuse}} = \frac{G}{Y} = \frac{10}{21.3}$$

$$= 0.469$$

$$\theta = \text{inverse cosine } 0.469$$

$$= 62°$$

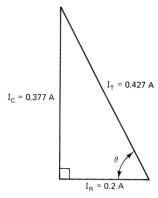

Current triangle

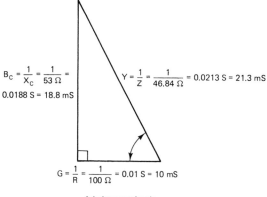

Admittance triangle

(c)

Fig. 3-17. Example of a parallel *RC* circuit. (a) Circuit. (b) Procedure for finding circuit values. (c) Circuit triangles.

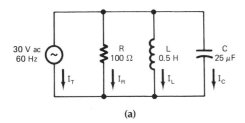

(a)

(b) Finding circuit values:

1. Find inductive reactance (X_L):

   $X_L = 2\pi \cdot f \cdot L$

   $= 6.28 \times 60 \times 0.5$

   $= 188.4\ \Omega$

2. Find capacitive reactance (X_C):

   $X_C = \dfrac{1{,}000{,}000}{6.28 \times 60 \times 25}$

   $= 106\ \Omega$

3. Find current through R (I_R):

   $I_R = \dfrac{V}{R} = \dfrac{30\ V}{100\ \Omega} = 0.3\ A$

4. Find current through L (I_L):

   $I_L = \dfrac{V}{X_L} = \dfrac{30\ V}{188.4\ \Omega} = 0.159\ A$

5. Find current through C (I_C):

   $I_C = \dfrac{V}{X_C} = \dfrac{30\ V}{106\ \Omega} = 0.283\ A$

6. Find total reactive current (I_X):

   $I_X = I_C - I_L$

   $= 0.283 - 0.159$

   $= 0.124\ A$

7. Find total current (I_T):

   $I_T = \sqrt{I_R^2 + I_X^2}$

   $= \sqrt{0.3^2 + 0.124^2}$

   $= \sqrt{0.9 + 0.0154}$

   $= \sqrt{0.1054}$

   $= 0.325\ A$

8. Check to see that I_T is larger than I_R or I_X.

9. Find impedance (Z):

   $Z = \dfrac{V}{I_T} = \dfrac{30\ V}{0.325\ A} = 92.3\ \Omega$

10. Check to see that Z is less than R or X_L or X_C.

11. Find circuit phase angle (θ):

    $\tan \theta = \dfrac{\text{opposite}}{\text{adjacent}} = \dfrac{I_X}{I_R} = \dfrac{0.124}{0.3}$

    $= 0.413$

    $\theta = \text{inverse tangent } 0.413$

    $= 22°$

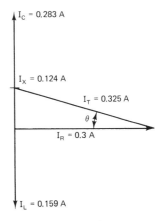

Current triangle

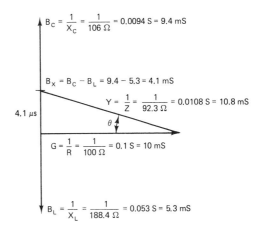

Admittance triangle

(c)

Fig. 3-18. Example of a parallel *RLC* circuit. (a) Circuit. (b) Procedure for finding circuit values. (c) Circuit triangles.

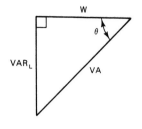

VA = volt-amperes (apparent power)
W = watts (true power)
VAR$_L$ = volt-amperes reactive (inductive)
VAR$_C$ = volt-amperes reactive (capacitive)
VAR$_T$ = volt-amperes reactive (total)

(a)

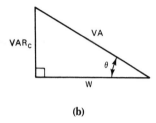

(b)

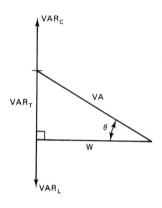

(c)

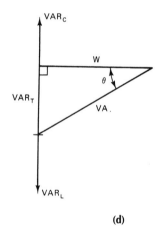

(d)

Calculating power values:

1. In dc circuits:

$$P = V \times I = I^2 \times R = \frac{V^2}{R}$$

2. In ac circuits:

a. $VA = V_A \times I_T = I_T^2 \times Z = \dfrac{V_A^2}{Z}$

b. $W = V_R \times I_R = I_R^2 \times R = \dfrac{V_R^2}{R}$

c. $VAR_L = V_L \cdot I_L = I_L^2 \cdot X_L = \dfrac{V_L^2}{X_L}$

d. $VAR_C = V_C \cdot I_C = I_C^2 \cdot X_C = \dfrac{V_C^2}{X_C}$

e. $VAR_T = VAR_L - VAR_C$ or $VAR_C - VAR_L$

(e)

Fig. 3-19. Power triangles for ac circuits. (a) *RL* circuits. (b) *RC* circuits. (c) *RLC* circuit with X_C larger than X_L. (d) *RLC* circuit with VAR_C larger than VAR_L. (e) Formulas for calculating power values.

applied to an ac circuit and current flow through the circuit. More inductive or capacitive reactance causes a larger phase angle. In a purely inductive or capacitive circuit, a 90° phase angle causes a power factor of 0°. The power factor varies according to the values of resistance and reactance in a circuit.

Two types of power affect power conversion in ac circuits. Power conversion in the resistive part of the circuit is called *active power*, or *true power*. True power is measured in watts (W). The other type of power is caused by inductive or capacitive loads. It is 90° out of phase with the true power and is called *reactive power*. It is a type of unused power. Reactive power is measured in volt-amperes reactive (VAR).

The power triangles of Fig. 3-19 have true power (W) marked as a horizontal line. Reactive power (VAR) is drawn at a 90° angle from the true power. Volt-amperes or apparent power (VA) is the longest side (hypotenuse) of the right triangle. This right triangle is similar to the impedance triangle and the voltage triangle for series ac circuits and the current triangle and admittance triangle for parallel ac circuits. Each type of right triangle has a *horizontal line* marking the *resistive* part of the circuit. A *vertical line* is used to mark the *reactive* part of the circuit. The *hypotenuse* is used to mark *total values* of the circuit. The length of the hypotenuse depends on the amount of resistance and reactance in the circuit. Vector diagrams and right triangles are extremely important for understanding ac circuits.

We can further examine the power relations of a power triangle by expressing each value mathematically on the basis of the value of apparent power (VA) and the phase angle (θ). The phase angle is the amount of phase shift, in degrees, between voltage and current in the circuit. Trigonometric ratios, which are discussed in appendix B, show that the sine of an angle of a right triangle is expressed as follows:

$$\text{sine } \theta = \frac{\text{opposite side}}{\text{hypotenuse}}$$

Because this is true, the phase angle can be expressed as follows:

$$\text{sine } \theta = \frac{\text{reactive power (VAR)}}{\text{apparent power (VA)}}$$

Therefore

$$\text{VAR} = \text{VA} \times \text{sine } \theta.$$

The phase angle and VAR can be determined with trigonometric ratios. The cosine of an angle of a right triangle is expressed as follows:

$$\text{cosine } \theta = \frac{\text{adjacent side}}{\text{hypotenuse}}$$

Thus in terms of the power triangle:

$$\text{cosine } \theta = \frac{\text{true power (watts)}}{\text{apparent power (VA)}}$$

Therefore true power can be expressed as follows:

$$W = VA \times \text{cosine}\,\theta$$

The following expression

$$\frac{\text{true power}}{\text{apparent power}}$$

is the power factor of a circuit; therefore the power factor is equal to the cosine of the phase angle (pf = cosine θ).

Self-Examination

Inductance and Inductive Reactance

Solve the following problems, which deal with inductance and inductive reactance:

1. Total inductance in series: $L_1 = 2$ H, $L_2 = 3$ H, $L_3 = 2$ H, $L_T = $ _____ H.
2. Total inductance in parallel: $L_1 = 2$ H, $L_2 = 3$ H, $L_3 = 8$ H, $L_T = $ _____ H.

Find the inductive reactance for the following:

3. 30 Hz, 8 H, $X_L = $ _____ Ω
4. 50 Hz, 2 H, $X_L = $ _____ Ω
5. 60 Hz, 1 H, $X_L = $ _____ Ω
6. 40 Hz, 3 H, $X_L = $ _____ Ω

Find the mutual inductance in series to increase inductance (field aiding):

7. $L_1 = 2$ H, $L_2 = 5$ H, $M = 0.55$

    $L_T = L_1 + L_2 + 2M = $ _____ H

8. $L_1 = 3$ H, $L_2 = 2$ H, $M = 0.35$

    $L_T = $ _____ H

Find the mutual inductance in series to decrease inductance (fields opposing):

9. $L_1 = 4$ H, $L_2 = 3$ H, $M = 0.85$

    $L_T = L_1 + L_2 - 2, M = $ _____ H

10. $L_1 = 3$ H, $L_2 = 3$ H, $M = 0.4$
    $L_T = $ _____ H

Compute the following problems when the inductors are connected in parallel. Use the values of the previous problem.

11. $\dfrac{1}{L_T} = \dfrac{1}{L_1 - M} + \dfrac{1}{L_2 - M}$, $L_T = $ _____ H

12. $L_T = $ _____ H

13. $\dfrac{1}{L_T} = \dfrac{1}{L_1 - M} + \dfrac{1}{L_2 - M}$, $L_T = $ _____ H

14. $L_T = $ _____ H

Capacitance and Capacitive Reactance

Solve the following problems, which deal with capacitance and capacitive reactance.

Compute the capacitance for the following capacitors, using the formula

$$C = \dfrac{0.0885ka}{t}$$

where C is capacitance in picofarads, k is the dielectric constant, a is the plate area in square centimeters, and t is the thickness of the dielectric in centimeters.

15. Dielectric constant = 3; area of plate = 2 × 200 cm; and thickness of dielectric = 0.03 cm:
    $C = $ _____ pF

16. Dielectric constant = 3; area of plate = 3 × 200 cm; and thickness of dielectric = 0.04 cm:
    $C = $ _____ pF

17. Dielectric constant = 2; area of plate = 0.25 × 275 cm; and thickness of dielectric = 0.02 cm:
    $C = $ _____ pF

Compute the following for the total capacitance in a series circuit using the following formula:

$$\dfrac{1}{C_T} = \dfrac{1}{C_2} + \dfrac{1}{C_2} + \dfrac{1}{C_3}$$

18. $C_1 = 10$ µF, $C_2 = 10$ µF, $C_3 = 30$ µF,
    $C_T = $ _____ µF

19. $C_1 = 40$ µF, $C_2 = 20$ µF, $C_3 = 20$ µF,
    $C_T = $ _____ µF

20. $C_1 = 80$ µF, $C_2 = 60$ µF, $C_3 = 80$ µF,
    $C_T = $ _____ µF

21. Compute the capacitive reactance for a 60 Hz ac circuit with a 30 µF capacitor: $X_C = $ _____ ohms.

Compute the following for total capacitance in parallel:

22. $C_1 = 80\ \mu F$, $C_2 = 60\ \mu F$, $C_3 = 80\ \mu F$,
    $C_T =$ _____ μF

23. $C_1 = 50\ \mu F$, $C_2 = 20\ \mu F$, $C_3 = 30\ \mu F$,
    $C_T =$ _____ μF

24. $C_1 = 70\ \mu F$, $C_2 = 50\ \mu F$, $C_3 = 50\ \mu F$,
    $C_T =$ _____ μF

Series and Parallel AC Circuits

Solve each of the following series and parallel ac circuit problems.

A 20 µF capacitor and a 1000 Ω resistor are connected in series with a 120 V, 60 Hz ac source. Find each of the following values:

25. $X_C =$ _____ Ω
26. True power = _____ W
27. $Z =$ _____ Ω
28. Apparent power = _____ VA
29. $I =$ _____ mA
30. Reactive power = _____ VAR
31. $V_C =$ _____ V
32. Power factor = _____ %
33. $V_R =$ _____ V

34. Draw an impedance triangle, a voltage triangle, and a power triangle for the values of Problems 25 through 33 and label each value.

A parallel ac circuit converts 12,000 W of power. The applied voltage is 240 V and the total current is 72 A. Find each of the following values:

35. Apparent power = _____ VA
36. Power factor = _____ %
37. Phase angle of circuit = _____ °

38. Draw a power triangle for the values of problems 35 through 37 and label each value.

A series circuit with 20 V, 60 Hz applied has a resistance of 100 Ω, a capacitance of 40 µF, and an inductance of 0.15 H. Find each of the following values.

39. $X_C =$ _____ Ω
40. $V_C =$ _____ V
41. $X_L =$ _____ Ω
42. $V_L =$ _____ V
43. $X_T =$ _____ Ω
44. $V_R =$ _____ V
45. $Z =$ _____ Ω
46. Phase angle = _____ °
47. $I =$ _____ mA

48. Draw an impedance triangle and a voltage triangle for the circuit of the previous problem and label each value.

Use the values given for R, C, and L in problems 39 through 47. Connect them in a parallel circuit that has 10 V applied to it.

Then find each value.

49. I_R = _____ mA   50. Conductance = _____ S
51. I_C = _____ mA   52. Susceptance = _____ S
53. I_L = _____ mA   54. Admittance = _____ S
55. I_T = _____ mA   56. Phase angle = _____ °
57. Z = _____ Ω

58. Draw a current triangle and an admittance triangle and label each value for the circuit of problems 49 through 57.

Complete the following sentences:

59. The property of an electric circuit that tends to prevent a change in current is called _____.

60. The letter symbol for inductance is _____.

61. Inductance is measured in _____.

62. A coil used to add inductance into an ac circuit is called a(n) _____.

63. The opposition to current flow caused by inductance is called _____.

64. The letter symbol for inductive reactance is _____.

65. Inductive reactance is measured in _____.

66. The total opposition to current flow in an ac circuit is called _____.

67. The letter symbol for impedance is _____.

68. Impedance is measured in _____.

69. The relation in electric degrees of voltage and current in an ac circuit is called _____.

70. In a purely inductive circuit, current _____ voltage by _____°.

71. In an ac circuit containing both resistance and inductance, current _____ voltage by _____ through _____°.

Resistance, Inductance, and Capacitance in AC Circuits 75

72. Maximum effective voltage multiplied by maximum effective current equals _____ power.

73. Apparent power is measured in _____.

74. The power consumed by the resistance in an ac is called the _____ power.

75. True power is measured in _____.

76. The ratio of true power to apparent power is called _____.

77. Power factor is expressed as a(n) _____ or a(n) _____.

78. Apparent power multiplied by power factor equals _____.

Answers

1. 7	2. 1.047
3. 1507.2	4. 628
5. 376.8	6. 753.6
7. 8.1	8. 5.7
9. 5.3	10. 5.2
11. 1.748	12. 1.38
13. 1.278	14. 1.3
15. 3540	16. 3982.5
17. 608.43	18. 4.29
19. 8 µF	20. 24.04
21. 88.46	22. 220
23. 100	24. 170
25. 132.7	26. 14.16
27. 14.16	28. 1008
29. 14.28	30. 119
31. 1.88	32. 99%
33. 119	34. $X_C \triangleleft \frac{Z}{R}$ $V_C \triangleleft \frac{V_A}{V_R}$ $V_{WC} \triangleleft \frac{V_A}{W}$
35. 17,280	36. 69%

Answers continued

37. 46.37
38. [power triangle diagram: W, θ, VA]
39. 66.35
40. 13.2
41. 56
42. 11.25
43. 9.83
44. 19.9
45. 100.48
46. 5.73°
47. 199
48. [impedance triangle: X_L, R, Z, θ; voltage triangle: V_L, V_R, V_A, θ]

49. 100
50. 0.01
51. 150
52. 0.0177
53. 177
54. 0.0104
55. 103.6
56. 15.15°
57. 96.5
58. [current and admittance phasor diagrams showing I_C, I_X, I_T, I_R, I_L and B_C, B_X, Y, G, B_L]

59. Inductance
60. L
61. Henrys
62. Inductor
63. Inductive reactance
64. X_L
65. Ohms
66. Impedance
67. Z
68. Ohms
69. Phase angle
70. Lags, 90°
71. Lags, 0–90°
72. Peak
73. Volt-amperes (VA)
74. True
75. Watts
76. Power factor
77. Decimal, percentage
78. True power

EXPERIMENT 3-1

INDUCTANCE AND INDUCTIVE REACTANCE

Inductance is an opposition to ac. It does not oppose dc. Because ac encounters many forms of opposition not encountered by dc, all quantities that oppose ac are called *impedance*. The letter Z in mathematical formulas represents these opposing quantities. One of the impedance quantities that oppose ac is *inductance*.

Inductance is the characteristic of an ac circuit that opposes any increase or decrease in current. Inductance is represented by L in mathematical formulas and is measured in henrys (H). Physically inductance is embodied in an inductor, which is no more than a coil. The impedance (Z) caused by an opposition other than resistance in ac circuits is known as *reactance*. In mathematical formulas, the letter X represents reactance. Reactance brought about by inductance is represented by X_L and is measured in ohms (Ω). The following formula is used to compute inductive reactance:

$$X_L = 2\pi f L$$

where

X_L = inductive reactance in ohms

2π = 6.28

f = frequency of ac in hertz

L = inductance in henrys

OBJECTIVES

1. To examine the effects of ac on inductive reactance.
2. To make computations concerning total inductance in series and parallel.

EQUIPMENT

VOM (multimeter)

Audio signal generator

Variable ac power source

Inductor: 4.5 H, iron core (or any value from 2 to 15 H)

Resistor: 10 kΩ

Switch: SPST

6 V lamp and socket

Connecting wires

PROCEDURE

1. Measure and record the resistance of the inductor and 6 V lamp filament.

    Inductor dc resistance = _____ Ω

    Lamp filament resistance = _____ Ω

2. Construct the circuit in Fig. 3-1A.

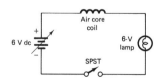

Fig. 3-1A. Series inductive dc circuit.

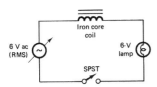

Fig. 3-1B. Series inductive ac circuit.

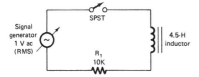

Fig. 3-1C. *RL* series circuit.

Ac frequency input	Voltage across R_1	Circuit current (I) $= \dfrac{V_{R_1}}{R_1}$	Inductive reactance (X_L) computed $= 2\pi \cdot f \cdot L$
100 Hz			
200 Hz			
300 Hz			
400 Hz			
500 Hz			
1 kHz			
10 kHz			

Fig. 3-1D. Characteristics of *RL* circuit.

3. Prepare the VOM to measure dc voltage. Close the switch, and measure and record the voltage across the lamp and coil.

   V (coil) = _____ V dc

   V (lamp) = _____ V dc

4. How does the sum of the voltages across the coil and the lamp compare with the source voltage? _____

   Why? _____

5. Using the data from steps 1 and 3, compute the current through the coil and lamp:

   I = _____ A.

6. Replace the dc power supply with the ac power source as shown in Fig. 3-1B.

7. Prepare the VOM to measure ac voltage. Close the SPST switch. Measure and record the voltage across the lamp and coil.

   V (lamp) = _____ V ac

   V (coil) = _____ V ac

8. How does the sum of these voltages compare with the source voltage? _____
   Why? _____

9. How does the sum of the voltages in step 4 compare with the sum of the voltages in step 8? _____

   Why was this relationship different? _____

10. Construct the circuit shown in Fig. 3-1C.

11. Prepare the VOM to measure ac voltage and connect it across R_1.

12. Close the switch and complete Fig. 3-1D. Adjust the signal generator to produce the ac frequencies indicated. (*Note:* The input voltage from the signal generator must be maintained at a constant level for each frequency.)

13. What is the relation between ac frequency and circuit current? _____

14. What is the relation between ac frequency and inductive reactance? _____

15. What is the relation between inductive reactance and circuit current? _____

Resistance, Inductance, and Capacitance in AC Circuits 79

ANALYSIS

1. What factors determine the inductance of an inductor? _____

2. What factors determine the value of inductive reactance? _____

3. Why does inductance cause ac current to lag voltage? _____

4. What is the total inductance when inductors of 4 H and 3 H are connected in series with no mutual inductance factor?

5. What is the total inductance when inductors of 4 H and 3 H are connected in parallel with no mutual inductance factor?

6. Assume the inductors in Question 4 are connected to aid with a mutual inductance of 0.6 H. What is the total inductance?

7. Assume the inductors in question 4 are connected to oppose with a mutual inductance of 0.86 H. What is the total inductance?

8. Assume the inductors in question 5 are connected to aid with a mutual inductance of 0.2 H. What is the total inductance?

9. Assume the inductors in question 5 are connected to oppose with a mutual inductance of 0.9 H. What is the total inductance?

10. What is mutual inductance?

11. If two inductors valued at 8 H and 10 H are connected in parallel, which allows the most ac current when the ac frequency is 1000 Hz?

12. Why will an inductor oppose ac more than dc for any given voltage?

EXPERIMENT 3-2

CAPACITANCE AND CAPACITIVE REACTANCE

Capacitance is the property of an ac circuit to oppose any increase or decrease in voltage. It is present any time two conductors or plates are separated by a dielectric material. Capacitance is represented by C in mathematical formulas and is measured in farads (F). Components designed to add capacitance to a circuit are known as *capacitors* and are rated according to their *capacitance* in farads and their *working voltage* (maximum dc voltage that can be placed across a capacitor without doing damage to its dielectric). Capacitors, known for their ability to store an electric charge, may be connected in series or parallel.

Capacitance in ac circuit causes current to *lead* the voltage. The impedance (Z) to ac caused by capacitance is known as *capacitive reactance* (X_C) and is measured in ohms (Ω). Capacitive reactance is computed with the following formula:

$$X_c = \frac{1}{2\pi f C}$$

where

X_C = capacitive reactance in ohms

2π = 6.28

f = frequency in hertz

C = capacitance in farads

Capacitors may be further classified as electrolytic or nonelectrolytic. If electrolytic capacitors are used in a circuit, the polarity printed on the body of the capacitor *must be observed*.

OBJECTIVES

1. To study capacitance and capacitive reactance.

2. To examine how ac affects capacitive reactance.

3. To make simple computations concerning total capacitance in series and parallel.

EQUIPMENT

VOM (multimeter)

Signal generator

6 V battery

Resistors: 10 kW

Capacitor: 0.01 µF, 100 V dc

Switch

Connecting wires

PROCEDURE

1. Prepare the VOM to measure resistance of approximately 1000 Ω or its equivalent. Measure and record the resistance of the 0.01 µF capacitor and the 10 kΩ resistor.

Resistor resistance = _____ Ω

Capacitor resistance = _____ Ω

2. How do these resistances compare?
   _____

3. From the data shown in step 1, which component would allow a dc to flow and which would not?
   Would: _____
   Would not: _____

4. Construct the circuit in Fig. 3-2A.

5. Prepare the VOM to measure dc of approximately 10 mA. Connect the VOM to the circuit to measure current. Close the Single Pole Single Throw (SPST) switch and record the current: _____ mA.

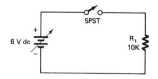

Fig. 3-2A. Simple resistive circuit.

6. Alter the circuit in step 4 to the circuit shown in Fig. 3-2B.

7. Connect the VOM to measure dc (in the 10-mA range). Close the switch and record the current: _____ mA.

8. How did the current recorded in step 5 compare with the current recorded in step 7?
   _____

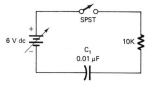

Fig. 3-2B. Simple RC dc circuit.

9. How do you account for the current difference in steps 5 and 7?
   _____

10. Alter the circuit in Fig. 3-2B so it is like the one in Fig. 3-2C.

11. Prepare the VOM to measure ac of approximately 10 V.

12. Connect the VOM across R_1, close the SPST switch, and record the voltage: V across R_1 = _____ V ac.

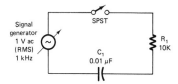

Fig. 3-2C. Simple RC ac circuit.

13. Using the resistance of R_1 as measured in step 1 and the voltage measured in step 12, compute the current in this circuit: I = _____ mA.

14. How did the current computed in step 13 compare with the current measured in step 7?
    _____

15. Because the two circuits are similar, how do you account for the difference in the currents in steps 7 and 13? _____

16. With the VOM connected as in step 12, adjust the signal generator to the frequencies shown in Fig. 3-2D. Complete the figure for each frequency.

Frequency (Hz)	Measured voltage across R_1 (V_{R_1})	Computed current $I = \dfrac{V_{R_1}}{R_1}$	Computed $X_c = \dfrac{1}{2\pi \cdot f \cdot C}$
300			
500			
700			
900			
1100			
1300			
1500			
1700			
1900			
2500			
3000			

Fig. 3-2D. Characteristics of RC circuit.

17. From the data recorded in Fig. 3-2D, what is the relation between frequency, current, and capacitive reactance? As frequency increases, the voltage across R_1 _____, current _____, and X_L _____.

ANALYSIS

1. What is capacitive reactance?

2. What is meant by a capacitor's working voltage?

3. What variables determine the capacitance of a capacitor?

4. If you connected in series two capacitors valued at 4 µF each what would be the total capacitance?

5. If you connected the two capacitors of question 4 in parallel, what would be their total capacitance?

6. What is the relation between ac frequency and X_C?

7. How does capacitance affect the phase relation between ac and voltage?

8. How does the relation between ac frequency and X_C compare with the relation between ac frequency and X_L?

EXPERIMENT 3-3

SERIES *RL* CIRCUITS

Series *RL* circuits are encountered very often in electronic equipment. When ac voltage is applied to this type of circuit, the current is the same through each component. However, the voltage that drops across each component is distributed according to the relative value of resistance (R) and inductive reactance (X_L) in the circuit.

The total opposition to current is called impedance (Z). Resistance and reactance in ac circuits both oppose the current. The impedance of an ac circuit is expressed as follows:

$$Z = \frac{V}{I}$$

or for a series *RLC* circuit:

$$Z = \sqrt{R^2 + (X_L - X_C)^2}$$

OBJECTIVE

To observe the characteristics of a series ac circuit that has resistance and inductance.

EQUIPMENT

AC voltage source

Resistor: 1 kΩ

Inductor: 4.5 H

VOM (multimeter)

PROCEDURE

1. Construct the series *RL* circuit shown in Fig. 3-3A. Apply 15 V ac from an ac power source.

2. Calculate the following values of the circuit:
   a. Inductive reactance: $X_L = 2\pi f L =$ _____ Ω
   b. Impedance: $Z = \sqrt{R^2 + X_L^2} =$ _____

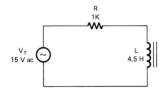

Fig. 3-3A. *RL* series circuit.

3. With a VOM, measure the following values:
   a. Voltage across the resistor:
      $V_R =$ _____ V ac
   b. Voltage across the inductor:
      $V_L =$ _____ V ac

4. If an ac milliammeter is available, measure the current in the circuit. If not, calculate $I_T = V_R/R$;
   $I_T =$ _____ mA

5. Complete the *impedance triangle* in Fig. 3-3B for the circuit you constructed by using the calculated values of R, X_L, and Z. The symbol θ represents the phase angle of the circuit.

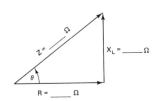

Fig. 3-3B. Impedance triangle for circuit.

6. Calculate the voltage drops in the circuit.
   a. $V_R = I_T \times R = $ _____ V ac
   b. $V_L = I_T \times X_L = $ _____ V ac

7. Show that the voltages are related in the following way by using the values you calculated and $V_T = 15$ V ac:

   $$V_T = \sqrt{V_R^2 + V_L^2} =$$

   _____ V ac.

8. Complete the *voltage triangles* for the circuit in Fig. 3-3C using the calculated and then the measured values of V_T, V_R, and V_L.

9. Calculate the power converted in the circuit:
   a. True power = $I_T \times V_R = $ _____ mW
   b. Apparent power = $I_T \times V_T = $ _____ mVA
   c. Reactive power = $I_T \times V_L = $ _____ mVAR

10. Complete the *power triangle* for the circuit in Fig. 3-3D using your calculated values of true power, apparent power, and reactive power.

ANALYSIS

1. Why must the voltages in a series *RL* circuit be added by using a right triangle?
   _____

2. Compare the calculated and measured values of the following:

Value	Calculated	Measured
I_T	_____	_____
V_R	_____	_____
V_L	_____	_____

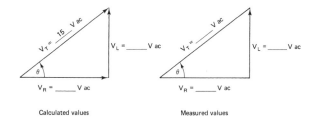

Fig. 3-3C. Voltage triangles for circuit.

Fig. 3-3D. Power triangle for circuit.

3. What are some factors that could cause a difference in your measured values and the calculated values? _____

4. Determine the phase angle of the circuit by using the following trigonometric methods.
    a. Phase angle (θ) = inverse tangent (X_L/R) = _____ °
    b. Phase angle (θ) = inverse tangent (V_L/V_R) = _____ °
    c. Phase angle (θ) = inverse tangent (VAR/W) = _____ °

5. Determine the power factor of the circuit as follows:

$$pf = \frac{\text{true power}}{\text{apparent power}} = \text{_____}$$

6. Determine the cosine of the phase angle (θ). This value should equal the circuit power factor (approximately pf = cosine θ): cosine θ = _____.

7. What is meant by power factor? _____

EXPERIMENT 3-4

SERIES RC CIRCUITS

Series *RC* circuits are used in many types of electronic equipment. When ac voltage is applied to this type of circuit, the characteristics are similar to those of a series *RL* circuit. In a capacitive circuit, *current leads voltage* and in an inductive circuit *voltage leads current*. Therefore these reactive values act in the opposite directions. In circuit calculations, the smaller reactance value is subtracted from the larger value to obtain the total reactance (X_T) of the circuit.

OBJECTIVE

To study the characteristics of a series ac circuit that has resistance and capacitance.

EQUIPMENT

Variable ac voltage source

Resistor: 100 Ω

Capacitor: 10 μF

VOM (multimeter)

PROCEDURE

1. Construct the series *RC* circuit shown in Fig. 3-4A. Apply 15 V ac from an ac power source.

2. Calculate the following values of the circuit:
   a. Capacitive reactance: $X_C = \dfrac{1}{2\pi f C} =$ _____ Ω
   b. Impedance: $Z = R^2 + X_C^2 =$ _____ Ω

3. With a VOM, measure the following values:
   a. Voltage across the resistor:
      $V_R =$ _____ V ac
   b. Voltage across the capacitor:
      $V_C =$ _____ V ac

4. If an ac milliammeter is available, measure the current in the circuit. If not, calculate the current as $I_T = V_R/R : I_T =$ _____ ac mA.

5. Complete the *impedance triangle* in Fig. 3-4B for the circuit you constructed by using the calculated values of R, X_C, and Z.

6. Calculate the voltage drops in the circuit as follows:
   a. $V_R = I_T \times R =$ _____ V ac
   b. $V_C = I_T \times X_C =$ _____ V ac

7. Show that the voltages are related in the following way by using the values you calculated:

$$V_T = \sqrt{V_R^2 + V_C^2} = \text{_____ V ac.}$$

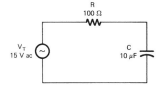

Fig. 3-4A. *RC* series circuit.

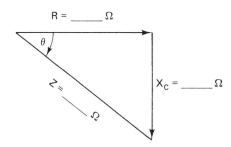

Fig. 3-4B. Impedance triangle for circuit.

8. Complete the voltage triangles in Fig. 3-4C for this circuit using the calculated then the measured values of V_T, V_R, and V_C.

9. Calculate the power converted in the circuit:
   a. True power = $I_T \times V_R$ = _____ mW
   b. Apparent power = $I_T \times V_C$ = _____ mVar

10. Complete the *power triangle* for the circuit in Fig. 3-4D by using your calculated values of true power, apparent power, and reactive power.

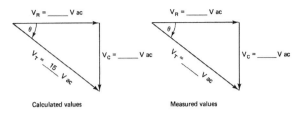

Fig. 3-4C. Voltage triangles for circuit.

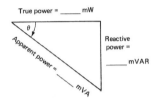

Fig. 3-4D. Power triangle for circuit.

ANALYSIS

1. Compare the calculated and measured values of the following:

Value	Calculated	Measured
I_T	_____ mA	_____ mA
V_R	_____ V	_____ V
V_C	_____ V	_____ V

2. Determine the phase angle of the circuit by using the following trigonometric methods:
   a. Phase angle (θ) = inverse tangent (X_C/R) = _____ °
   b. Phase angle (θ) = inverse tangent (V_C/V_R) = _____ °
   c. Phase angle (θ) = inverse tangent (W/VA) = _____ °

3. Determine the power factor of the circuit as:

   $$pf = \frac{\text{true power}}{\text{apparent power}} = \underline{\qquad}$$

4. Determine the cosine of the phase angle. This value should equal the power factor of the circuit: cosine θ = _____ .

88 UNIT 3

EXPERIMENT 3-5

SERIES *RLC* CIRCUITS

In this experiment you will study the characteristics of an ac circuit that has resistive, capacitive, and inductive components. Because the effects of the capacitance and inductance are 180° out of phase with each other, you must analyze these circuits by using right triangles to show the relations of the values.

OBJECTIVE

To study the characteristics of a series *RLC* circuit.

EQUIPMENT

AC voltage source

Resistor: 100 Ω

Capacitor: 10 μF

Inductor: 4.5 H

VOM (multimeter)

PROCEDURE

1. Construct the series *RLC* circuit shown in Fig. 3-5A. Apply 15 V ac from an ac power source.

2. Calculate the following:

   a. Capacitive reactance:
   $$X_C = \frac{1}{2\pi f C} = \underline{\qquad} \Omega$$

   b. Inductance reactance:
   $$X_L = 2\pi f L \underline{\qquad} \Omega$$

   c. Total reactance:
   $$X_T = X_L - X_C = \underline{\qquad} \Omega$$

   d. Impedance:
   $$Z = \sqrt{R^2 + (X_L - X_C)^2} = \underline{\qquad} \Omega$$

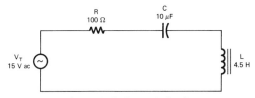

Fig. 3-5A. Series *RLC* circuit.

3. With a VOM measure the following values:

   a. Voltage across the capacitor:
   $V_C = \underline{\qquad}$ V ac

   b. Voltage across the inductor:
   $V_L = \underline{\qquad}$ V ac

   c. Voltage across the resistor:
   $V_R = \underline{\qquad}$ V ac

4. If an ac milliammeter is available, measure the current through the circuit. If not, calculate as $I_T = V_R/R$: $I_T = \underline{\qquad}$ mA.

5. Complete the *impedance triangle* in Fig. 3-5B for the circuit you constructed by using the calculated values of R, X_C, X_L, and Z.

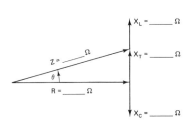

Fig. 3-5B. Impedance triangle for circuit.

6. Calculate the voltage drops in the circuit as follows:
   a. $V_R = I_T \times R =$ _____ V ac
   b. $V_C = I_T \times X_C =$ _____ V ac
   c. $V_L = I_T \times X_L =$ _____ V ac
   d. $V_X = V_L - V_C =$ _____ V ac

7. Show that these voltages are related in the following way by using the values you calculated:

   $$V_T = \sqrt{V_R^2 + (V_L - V_C)^2} = \text{_____} \text{ V ac}$$

8. Complete the voltage triangles in Fig. 3-5C for this circuit using the calculated and then the measured values and V_T, V_R, V_L, and V_C.

9. Calculate the power converted in the circuit:
   a. True power: $I_T \times V_R =$ _____ W
   b. Apparent power: $I_T \times V_T =$ _____ mVA
   c. Capacitive reactive power: $VAR_C = I_T \times V_L =$ _____ mVAR
   d. Inductive reactive power: $VAR_L = I_T \times V_L =$ _____ mVAR
   e. Total reactive power: $VAR_T = I_T \times V_X =$ _____ mVAR

10. Complete the *power triangle* in Fig. 3-5D for the circuit by using your calculated values of true power, apparent power, and reactive power.

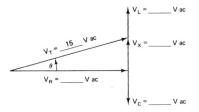

Calculated values

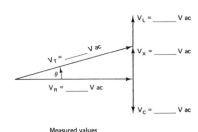

Measured values

Fig. 3-5C. Voltage triangles for circuit.

ANALYSIS

1. Compare the calculated and measured values of the following:

Value	Calculated	Measured
I_T	_____ mA	_____ mA
V_R	_____ V	_____ V
V_C	_____ V	_____ V
V_L	_____ V	_____ V

2. Determine the phase angle of the circuit by using the following trigonometric methods:
   a. Phase angle (θ) = inverse tangent (X_T/R) = _____ °
   b. Phase angle (θ) = inverse tangent (V_X/V_R) = _____ °
   c. Phase angle (θ) = inverse tangent (VAR_T/W) = _____ °

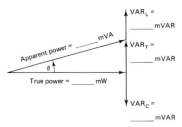

Fig. 3-5D. Power triangle for circuit.

3. Determine the power factor of the circuit as follows:

$$pf = \frac{\text{true power}}{\text{apparent power}} = \underline{\hspace{2cm}}$$

4. Determine the cosine of the phase angle. This value should equal the circuit power factor:
cosine θ = _____

5. How does the circuit of this experiment show power factor correction (reduction of phase angle by adding capacitance?) _____

6. As total reactive power increases, the phase angle _____.

EXPERIMENT 3-6

PARALLEL *RL* CIRCUITS

Another category of basic ac circuits is the parallel type. Parallel circuits have many applications in electronics. The basic calculations necessary to study parallel ac circuits are somewhat different from those used to study series circuits. Because the impedance (Z) of a parallel circuit is less than individual branch values of resistance, inductive reactance, or capacitive reactance, the impedance relations must be modified. A right-triangle relation exists between the currents in the branches of the circuit.

OBJECTIVE

To study the impedance and current relations of a parallel resistive inductive (*RL*) circuit with ac applied.

EQUIPMENT

AC voltage source

Resistor: 100 Ω

Inductor: 4.5 H

VOM (multimeter)

PROCEDURE

1. Construct the parallel *RL* circuit shown in Fig. 3-6A. Apply 15 V ac from an ac power source.

2. Calculate the following:

    a. Inductive reactance: $X_L = 2\pi f L =$ _____ Ω

    b. Current through the resistor: $I_R = V_R/R =$ _____ mA

    c. Current through the inductor: $I_L = V_L/X_L =$ _____ mA

    d. Total current:

    $$I_T = \sqrt{I_R^2 + I_L^2} = \underline{\quad\quad} \text{mA}$$

    e. Impedance: $Z = V_T/I_T =$ _____ Ω

3. Complete the current triangle (Fig. 3-6B) for this circuit using the values from step 2.

4. If an ac milliammeter is available, measure the following values. If not, calculate as follows:

    a. Current through the resistor: $I_R = V_T/R =$ _____ mA

    b. Current through the inductor: $I_L = V_T/X_L =$ _____ mA

    c. Total current: $I_T = V_T/Z =$ _____ mA

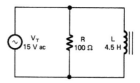

Fig. 3-6A. Parallel *RL* circuit.

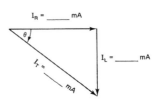

Fig. 3-6B. Current triangle for circuit.

5. Because the impedance is less than the individual value of resistance or inductive reactance of the circuit, an *admittance diagram* must be used. In this diagram, admittance $Y = 1/Z$, conductance $G = 1/R$, and inductive susceptance $B_L = 1/X_L$. The unit of measurement is the siemen.

6. Calculate the values necessary to complete the admittance triangle:
   a. Admittance: $Y = 1/Z =$ _____ µS
   b. Conductance: $G = 1/R =$ _____ µS
   c. Inductive susceptance: $B_L = 1/X_L =$ _____ µS

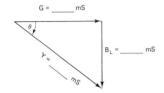

Fig. 3-6C. Admittance triangle for circuit.

7. Complete the *admittance triangle* (Fig. 3-6C) for the circuit.

8. Calculate the power converted in the circuit.
   a. True power:
   $$I_R^2 \times R = \text{_____} \text{ mW}$$
   b. Apparent power:
   $$I_T^2 \times Z = \text{_____} \text{ mVA}$$
   c. Reactive power:
   $$I_L^2 \times X_L = \text{_____} \text{ mVar}$$

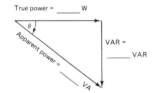

Fig. 3-6D. Power triangle for circuit.

9. Complete the *power triangle* (Fig. 3-6D) for the circuit.

ANALYSIS

1. Compare the calculated and measured values of the following:

Value	Calculated	Measured
I_R	_____ mA	_____ mA
I_L	_____ mA	_____ mA
I_T	_____ mA	_____ mA

2. Determine the phase angle of the circuit by the following methods:
   a. Phase angle: $\theta =$ inverse cosine $(I_R/I_T) =$ _____ °
   b. Phase angle: $\theta =$ inverse sine (VAR/VA) = _____ °

3. Determine the power factor of this circuit:
   $$pf = \frac{\text{true power}}{\text{apparent power}} = \text{_____}$$

Resistance, Inductance, and Capacitance in AC Circuits

4. Compare the power factor value of question 3 with the value of cosine θ from the current triangle of step 3 in the procedure. ―――――

5. Define the following terms associated with parallel ac circuits.
   a. Admittance ―――――――――――――――

   b. Conductance ―――――――――――――――

   c. Inductive susceptance ―――――――――

   d. Capacitive susceptance ―――――――――

EXPERIMENT 3-7

PARALLEL *RC* CIRCUITS

In this experiment you will study the characteristics of a parallel *RC* circuit with ac applied. This type of circuit is similar to the *RL* circuit; however, the effect of capacitance is opposite that of inductance. Similar triangle relations exist in this circuit.

OBJECTIVE

To study the impedance, current, and power relations of a parallel *RC* circuit.

EQUIPMENT

Variable ac voltage source

Resistor: 100 Ω

Capacitor: 10 μF

VOM (multimeter)

PROCEDURE

1. Construct the parallel *RC* circuit shown in Fig. 3-7A. Apply 15 V ac from an ac power source.

2. Calculate the following:

   a. Capacitive reactance:

   $$X_C = \frac{1}{2\pi f C} = \underline{\qquad} \, \Omega$$

   b. Current through the resistor: $I_R = V_R/R = \underline{\qquad}$ mA

   c. Current through the capacitor: $I_C = V_R/X_C = \underline{\qquad}$ mA

   d. Total current:

   $$I_T = \sqrt{I_R^2 + I_C^2} = \underline{\qquad} \text{ mA}$$

   e. Impedance: $Z = V_S/I_T = \underline{\qquad}$ Ω

3. Complete the *current triangle* (Fig. 3-7B) for this circuit using the values from step 2.

4. Measure the following ac current values with an ac milliammeter, if available, or calculate them as shown:

   a. Current through the resistor: $I_R = V_T/R = \underline{\qquad}$ mA

   b. Current through the capacitor: $I_C = V_T/X_C = \underline{\qquad}$ mA

   c. Total current: $I_T = V_T/Z = \underline{\qquad}$ mA

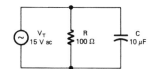

Fig. 3-7A. Parallel *RC* circuit.

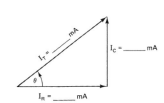

Fig. 3-7B. Current triangle for circuit.

5. Calculate the following values:
   a. Admittance: $Y = 1/Z =$ ———— µS
   b. Conductance: $G = 1/R =$ ———— µS
   c. Capacitive susceptance: $(B_C) = 1/X_C =$ ———— µS

6. Complete the *admittance triangle* (Fig. 3-7C) for the circuit.

7. Calculate the power converted in the circuit:
   a. True power:

   $$I_R^2 \times R = \text{————— mW}$$

   b. Apparent power:

   $$I_T^2 \times Z = \text{————— mVA}$$

   c. Reactive power:

   $$I_C^2 \times X_C = \text{————— mVar}$$

8. Complete the power triangle for the circuit in Fig. 3-7D.

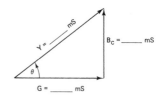

Fig. 3-7C. Admittance triangle for circuit.

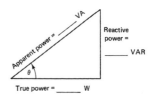

Fig. 3-7D. Power triangle for circuit.

ANALYSIS

1. Compare the calculated and measured values of the following:

Value	Calculated	Measured
I_R	———— mA	———— mA
I_C	———— mA	———— mA
I_r	———— mA	———— mA

2. Determine the phase angle (θ) of the circuit using the following methods:
   a. Phase angle: $\theta = $ inverse sine $(I_C/I_R) = $ ————°
   b. Phase angle: $\theta = $ inverse tangent (VAR/W) = ————°

3. What is the value of the following:
   a. Cosine $\theta = $ ———— (use correct triangle values)
   b. $pf = \dfrac{\text{true power}}{\text{apparent power}} = $ ————

4. Suppose that a 5 H inductor has been connected in parallel with the circuit of this activity, and compute the following:
   a. $X_L = 2\pi f L =$ _____ Ω
   b. $I_L = V_L / X_L =$ _____ mA
   c. $I_X = I_L - I_C =$ _____ mA
   d. $I_T \sqrt{I_R^2 + I_X^2} =$ _____ mA
   e. $Z = V_S / I_T =$ _____ Ω
   f. $\theta =$ inverse sine $(I_X / I_T) =$ _____ °
   g. pf = cosine $\theta =$ _____

5. How do the values of power factor for the original circuit and that of question 4 differ? _____

   Why? _____

6. How do the values of total current differ in the two circuits? _____
   Why? _____

Unit 3 Examination

Inductance and Capacitance in AC Circuits

Instructions: For each of the following, circle the answer that most correctly completes the statement.

1. Vectors indicate:
   a. Direction
   b. Magnitude
   c. Time, direction, and magnitude
   d. Magnitude and direction

2. Inductance is the property of an electric circuit that
   a. Opposes change in applied voltage
   b. Opposes change in current
   c. Aids changes in voltage
   d. Aids changes in current

3. Capacitance is the property of an electric circuit that
   a. Opposes any change in current in the circuit
   b. Opposes any change in voltage in the circuit
   c. Is not affected by a change in voltage
   d. Aids any change of current in the circuit

4. The unit of measurement of reactive power is:
   a. Volt-ampere (VA)
   b. Volt-ampere-wattage (VAW)
   c. Volt-ampere-kilowatt (VAkW)
   d. Volt-ampere-reactive (VAR)

5. The prime factors that determine inductive reactance are:
   a. Frequency and current
   b. Frequency and induced voltage
   c. Frequency and inductance
   d. Inductance and current

6. Assume that the component values in a series RCL circuit are $R = 4\ \Omega$, $X_C = 3\ \Omega$, and $X_L = 6\ \Omega$. The total reactance in the circuit is approximately
   a. $3\ \Omega$ b. $6\ \Omega$
   b. $8\ \Omega$ d. $9\ \Omega$

7. Assume that a series circuit contains both inductance and resistance. The phase angle between the applied voltage (E_A) and the resistive voltage drop (E_R) is the same as the phase angle between
   a. Impedance and reactance
   b. Resistance and impedance
   c. Reactance and resistance
   d. Resistance and susceptance

8. When frequency is increased in a series RL circuit, the current flow will
   a. Increase because of greater X_L
   b. Increase because of less X_L
   c. Decrease because of greater X_L
   d. Decrease because of less X_L

9. Voltage and current are considered to be out of phase with each other in a purely inductive circuit by what amount?
   a. Current leads voltage by 90°
   b. Current lags voltage by 90°
   c. Current leads voltage by 180°
   d. Current lags voltage by 180°

10. The phase angle between current and voltage in a circuit containing both resistive and inductive elements is:
    a. Greater than 0° but less than 90°
    b. A constant 45°
    c. 90° at all times
    d. 0° because X_L and R are equal

11. What is impedance?
    a. Opposition to the current flow in an ac circuit created by resistance and reactance
    b. Resistance of an ac circuit
    c. Vector sum of voltage drops in an ac circuit
    d. Current divided by voltage in an ac circuit

12. In a circuit containing both X_C and X_L, if the difference between X_C and X_L increases
    a. Total impedance decreases
    b. Total impedance increases
    c. Total impedance remains the same
    d. Resistance increases

13. The hypotenuse of the power triangle represents
    a. True power   b. VAR
    b. Apparent power   d. None of the above

14. The inductance of a coil is measured in which unit?
    a. Ohms
    b. Farads
    c. Henrys
    d. Volts

15. Inductive reactance in an ac circuit increases with
    a. An increase in frequency
    b. A decrease in resistance
    c. An increase in resistance
    d. A decrease in frequency

16. When the impedance of an ac circuit is greater than the dc resistance because of the presence of an inductor, the cause is
    a. X_C (capacitive reactance)
    b. X_R (resistive reactance)
    c. X_L (inductive reactance)
    d. X_M (mutual reactance)

17. In an ac circuit that contains an inductive device, the current
    a. Is in phase with the voltage
    b. Lags the voltage
    c. Leads the voltage
    d. Is 180° out of phase with the voltage

18. In an ac circuit that contains a capacitive device, the current
    a. Is in phase with voltage
    b. Lags the voltage
    c. Leads the voltage
    d. Is 180° out of phase with the voltage

19. Increasing the frequency of the applied ac in a capacitive circuit will cause capacitive reactance to
    a. Increase
    b. Decrease
    c. Remain constant
    d. Be increased to a resonant state

20. The capacitance of a capacitor is measured in
    a. Farads
    b. Henrys
    c. Ohms
    d. Mhos

21. Measured electric power expressed in watts is referred to as
    a. Apparent power
    b. True power
    c. Unity power
    d. Power factor

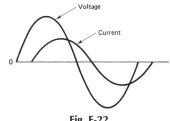

Fig. E-22

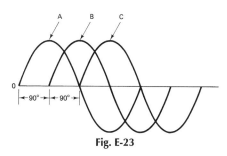

Fig. E-23

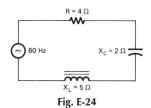

Fig. E-24

22. The waveforms in Fig. E-22 represent a power factor of
    a. 0   b. Cannot be determined
    c. 1.0   d. 0.5

23. The waveforms in Fig. E-22 are typical of a:
    a. Capacitive circuit   b. Inductive circuit
    c. Resistive circuit   d. Combination

24. The waveform in Fig. E-23 illustrates
    a. Three-phase ac   b. A capacitive circuit
    c. An inductive circuit   d. Two-phase ac

25. What is the impedance of the circuit of Fig. E-24?
    a. 25 Ω   b. 50 Ω
    c. 11 Ω   d. 5 Ω

UNIT 4

Transformers

Unit Introduction

Transformers are important electronic devices. They are used either to increase or to decrease alternating current (ac) voltage. Transformers are made with two separate sets of wire windings on a metal core. These are called the *primary winding* and the *secondary winding*. A transformer that increases voltage is called a *step-up transformer*; one that decreases voltage is called a *step-down transformer*.

UNIT OBJECTIVES

Upon completion of this unit, you should be able to:

1. List several purposes of transformers.
2. Describe the construction of a transformer.
3. Explain transformer action.
4. Calculate turns ratio, voltage ratio, current ratio, power, and efficiency of transformers.
5. Explain the purpose of isolation transformers and autotransformers.
6. Explain factors that cause losses in transformer efficiency.
7. Investigate the characteristics of transformers.

Important Terms

Before studying unit 4, you should review the following terms.

Center tap A terminal connection made to the center of a transformer winding.

Isolation transformer A transformer with a 1:1 turns ratio used to isolate an ac power line from equipment with a chassis ground.

Mutual inductance (M) The condition in which two coils are located close together so that the magnetic fluxes of the coils affect one another in terms of inductance properties.

Primary winding The coil of a transformer to which ac source voltage is applied.

Secondary winding The coil of a transformer into which voltage is induced; energy is delivered to the load circuit through the secondary winding.

Step-down transformer A transformer in which the secondary voltage is lower than the primary voltage.

Step-up transformer A transformer in which the secondary voltage is higher than the primary voltage.

Transformer An ac power control device that transfers energy from its primary winding to its secondary winding by means of mutual inductance and is ordinarily used to increase or decrease voltage.

Turns ratio The ratio of the number of turns of the primary winding (N_p) of a transformer to the number of turns of the secondary winding (N_s).

Transformer Operation

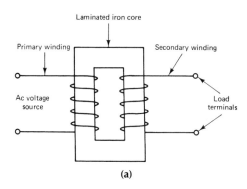

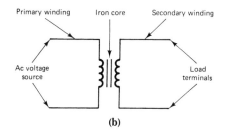

Fig. 4-1. Transformer: (a) pictorial; (b) schematic symbol.

Transformers are electrical control devices used either to increase or to decrease ac voltage. They do not operate with dc voltage applied. Figure 4-1 shows ac voltage applied to the primary winding of the transformer. There is no connection of the primary and secondary windings. The transfer of energy from the primary to the secondary winding is due to magnetic coupling or mutual inductance. The transformer relies on electromagnetism to operate.

The primary and secondary windings of transformers are wound around a laminated iron core. The iron core is used to transfer the magnetic energy from the primary winding to the secondary winding. AC voltage is applied to the primary winding; the secondary winding is connected to an electrical load.

There are many different types and sizes of transformers. However, the same basic principles of operation apply to all of them. The operation of a transformer relies on the expanding and collapsing of the magnetic field around the primary winding. When current flows through a conductor, a magnetic field develops around the conductor. Application of AC voltage to the primary winding causes a constantly changing magnetic field around the winding. During times of increasing ac voltage, the magnetic field around the primary winding expands. After the peak value of the ac cycle is reached, the voltage decreases toward zero. When ac voltage decreases, the magnetic field around the primary winding collapses. The collapsing magnetic field is transferred to the secondary winding. Transformers will not operate with dc voltage applied because dc voltage does not change in value.

Types of Transformers

There are three basic transformer types—*air-core, iron-core,* and *powered metal-core transformers*. The core material used to construct transformers depends on the application of the transformers. Radios and televisions are types of air-core and powdered metal-core transformers. Iron-core transformers are sometimes called *power transformers*. They are often used in the power supplies of electrical equipment and at electrical power distribution substations.

Two other classifications of transformers are step-up transformers and step-down transformers. *Step-up transformers* have more turns of wire used for their secondary windings than for their primary windings. *Step-down transformers* have more turns of wire on their primary windings than on their secondary windings. A step-up transformer is shown in Fig. 4-2a. A step-up transformer has a higher voltage across the secondary winding than the input voltage applied to the primary winding. The step up is caused by the turns ratio of the primary and secondary windings. The *turns ratio* is the ratio of primary turns of wire (N_p) to secondary turns of wire (N_s). If 100 turns of wire are used for the primary winding and 200 turns for the secondary winding, the turns ratio is 1:2. The secondary voltage is twice the primary voltage. If 10 V ac is applied to the primary winding, 20 V ac develops across the load connected across the secondary winding. The voltage ratio (V_p/V_s) is the same as the turns ratio.

Figure 4-2b shows a step-down transformer. The primary winding has 500 turns and the secondary has 100 turns of wire. The turns ratio (N_p/N_s) is 5:1. This means that the primary voltage (V_p) is five times the secondary voltage (V_s). If 25 V is applied to the primary winding, 5 V develops across the secondary. A step-up transformer increases voltage, and a step-down decreases voltage. The voltage formula in Fig. 4-2 shows the relation between the primary and secondary voltages and the turns ratio of the primary and secondary windings.

Transformers are power-control devices. Power equals voltage (V) multiplied by current (I). In a transformer, primary power (P_p) equals secondary power (P_s). Therefore primary voltage (V_p) times primary current (I_p) equals secondary voltage (V_s) times secondary current (I_s), as follows:

$$P_p = P_s$$
or
$$V_p \times I_p = V_s \times I_s$$

The current ratio of transformers is an inverse ratio: $N_p/N_s = I_s/I_p$. Application of the current ratio is shown in Fig. 4-3. If voltage is increased, current decreases in the same proportion. The secondary voltage decreases 10 times and secondary current increases 10 times in the example shown. This is a step-down transformer because voltage decreased.

Multiple-Secondary Transformers

Some transformers have more than one secondary winding. These are called *multiple-secondary transformers*. Some are step-up and some are step-down transformers depending on the application. There are many other special types of transformers. *Autotransformers* have only one winding. They may be either step-up or step-down transformers depending on how they are connected. Variable autotransformers are used in power supplies to provide variable ac voltages.

Current Transformers

Current transformers are used to measure large ac currents. A conductor is placed inside the hole in the center. A current-carrying conductor produces a magnetic field. This magnetic

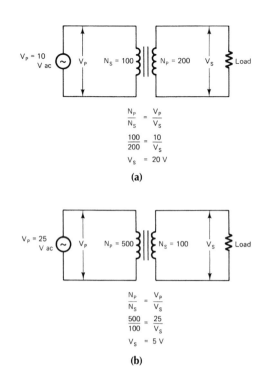

Fig. 4-2. Transformers: (a) step-up; (b) step-down.

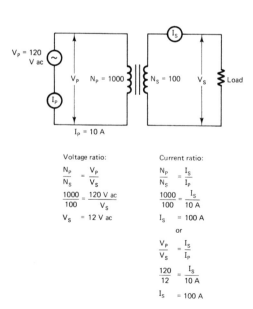

Fig. 4-3. Transformer current ratio.

field is transferred to the secondary winding of the current transformer. The transformer reduces the high current so that it can be measured.

Transformer Applications

Many transformers are used to control three-phase voltages. Large three-phase transformers are used at electrical power-distribution substations to step down voltages. Another type of transformer is called an *isolation transformer*. This type of transformer is used for tests with any equipment that plugs into an ac power outlet. Its purpose is to isolate the power source of the test equipment from the equipment being tested. An isolation transformer is plugged into a power outlet. The equipment being tested is plugged into the secondary winding of the isolation transformer. There is no connection between the primary and secondary windings of the transformer. This isolates the hot and neutral wires of the two pieces of equipment. The possibility of electrical hot to ground shorts is eliminated when an isolation transformer is used.

Transformer Efficiency

Transformers are very efficient electrical devices. A typical efficiency rating for a transformer is approximately 98%. The efficiency of electrical equipment is determined with the following formula:

$$Efficiency\ (\%) = \frac{P_{out}}{P_{in}} \times 100$$

where P_{out} is the secondary power ($V_s \times I_s$) in watts (W) and P_{in} is the primary power ($V_p \times I_p$) in watts.

Self-Examination

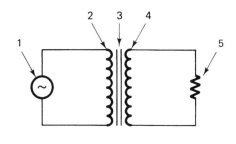

Fig. 4-4.

1.–5. Identify the parts of the transformer shown in Fig. 4-4. 1 _____; 2 _____; 3 _____; 4 _____; 5 _____

6. What is the secondary voltage of the transformer shown in Fig. 4-5? $V_a =$ _____.

7. If the primary current of the transformer shown in Fig. 4-5 is 5 A, what would the secondary current equal? $I_s =$ _____.

8. A transformer that has only one winding is called _____.

9. The main reason for using transformers to increase voltage for power transmission is to reduce _____.

10. An electrical device consisting of two coils very close together yet electrically insulated from each other is a _____.

11. A transformer coil connected to a source of ac voltage is the _____ winding.

12. A transformer coil connected to the load is the _____ winding.

13. If there are fewer turns in the secondary winding than in the primary winding, the device is a _____ transformer.

14. If there are more turns in the secondary winding than in the primary winding, the device is a _____ transformer.

15. To reduce opposition to magnetic lines of force between the primary and secondary windings, the windings may be wound on a(n) _____.

16. If 120 V is applied to the primary winding of a step-down transformer with a 10:1 turns ratio, the voltage induced in the secondary winding is _____ V.

17. If 120 V is applied to the primary winding of a step-up transformer with a 1:2 turns ratio, the voltage induced in the secondary winding is _____ V.

18. If the voltage across the complete secondary winding of a center-tap transformer is 240 V, the voltage from one outside conductor to the center tap is _____ V.

19. A transformer with more than one secondary winding is called a _____ transformer.

20. The ratio of power output to power input of a transformer is called _____.

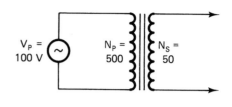

Fig. 4-5.

ANSWERS

1. AC source	2. Primary winding
3. Iron core	4. Secondary winding
5. Load	6. 10 V
7. 50 A	8. Autotransformer
9. Losses	10. Transformer
11. Primary	12. Secondary
13. Step-down	14. Step-up
15. Iron core	16. 12 V
17. 240 V	18. 120 V
19. Multiple-secondary	20. Efficiency

EXPERIMENT 4-1

TRANSFORMER ANALYSIS

In this activity, you may use any multiple-secondary transformer to complete an analysis of its voltage and resistance characteristics. A multiple-secondary transformer has more than one secondary winding. You should record the voltage and current ratings of each winding. Be careful in taking voltage measurements, because some secondary windings may be step-up windings, whereas others may be step-down windings. Also be careful not to short any of the secondary windings together because this may destroy the transformer.

OBJECTIVE

To study the voltage ratio, current ratio, and winding resistance characteristics of a multiple-secondary transformer.

EQUIPMENT

VOM (multimeter)
Multiple-secondary transformer
Variable ac source

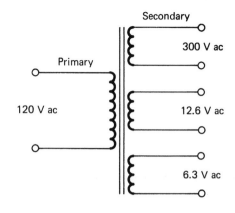

Fig. 4-1A.

PROCEDURE

1. Figure 4-1A shows a typical schematic for a multiple-secondary transformer. The transformer you use should be similar to this.

2. Record the voltages and current ratings of the secondary windings, which may be obtained by applying the rated voltage to the primary winding. These are the rated values for this particular transformer. Also record whether the secondary windings are step-up or step-down.

3. In the space below, draw a schematic diagram of the transformer you are using. Label the voltage and current rating of each winding if this information is available.

4. Apply the rated voltage to the primary winding and measure the voltage (V_s) across each secondary winding. Record these values:

   V_{s1} = _____ V ac; V_{s2} = _____ V ac;
   V_{s3} = _____ V ac; V_{s4} = _____ V ac;
   V_{s5} = _____ V ac

5. Reduce the applied voltage to one-half its rated value and complete the same secondary measurements as in step 4.

   V_{s1} = _____ V ac; V_{s2} = _____ V ac;
   V_{s3} = _____ V ac; V_{s4} = _____ V ac;
   V_{s5} = _____ V ac

6. With no power applied, make the following resistance measurements.

   Primary Resistance (R_p) = _____

   R_{s1} = _____ Ω

   R_{s2} = _____ Ω

   R_{s3} = _____ Ω

   R_{s4} = _____ Ω

   R_{s5} = _____ Ω

ANALYSIS

1. What conclusion may be drawn from the resistance measurements of step 6 concerning current capacity for each secondary winding?
   _____
   _____

2. What is the relation between primary power and secondary power in a multiple-secondary transformer? _____
   _____

3. Do the measured voltage values across the secondary windings correspond well with the rated values? _____ Give an example to justify your answer. _____
   _____

4. Discuss the following characteristics of transformers:

   a. Winding resistance _____
   _____

   b. Volt-ampere rating _____
   _____

   c. Core construction _____
   _____

   d. Magnetic coupling _____
   _____

   e. Turns ratio _____
   _____

   f. Voltage ratio _____
   _____

5. Compare the voltage measurements of steps 4 and 5. _____
   _____

Unit 4 Examination

Transformers

Instructions: For each of the following, circle the answer that most correctly completes the statement.

1. A transformer with more turns on the primary side than on the secondary side is called

    a. A flexiformer
    b. An autotransformer
    c. A step-up transformer
    d. A step-down transformer

2. A transformer that has only one winding is known as

    a. A flexiformer
    b. An autotransformer
    c. A step-up transformer
    d. A step-down transformer

3. Transformers are used in electrical power distribution to reduce

    a. Voltage changes   b. Resistance of wires
    c. Line loss   d. Power factor

4. For transformer action to take place, the primary and secondary circuits must be complete, and the transformer must have

    a. Two or more closely spaced windings
    b. A changing magnetic field
    c. A tapped winding
    d. A closed core

5. A comparison of step-up and step-down autotransformers indicates that the step-down transformer

    a. Has more turns in the primary than in the secondary
    b. Has fewer turns in the primary than in the secondary
    c. Requires more input voltage to operate
    d. Requires less input voltage to operate

6. Assume that the primary power input to a transformer with 95% efficiency is 100 W. The secondary output power of the transformer is

    a. 5 W   b. 90 W
    c. 95 W   d. 105 W

7. The amount of coupling of a transformer is
   a. Increased by turning one coil at a right angle to the other
   b. Increased with the addition of a soft iron core
   c. Higher in an air-core transformer than an iron-core type
   d. Almost unity for air-core types

8. The principle of operation of a transformer is
   a. Electromagnetic induction
   b. Varying a conductor in a magnetic field
   c. Mutual induction
   d. Thermionic emission

9. An ac distribution system that produces 120/208 V is
   a. A delta system
   b. A single-phase system
   c. A wye system
   d. A combination system

10. In a transformer with a small secondary load, secondary voltage
    a. Is 90° out of phase with the applied voltage
    b. Leads primary current by 90°
    c. Is in phase with secondary current
    d. Is 180° out of phase with the applied voltage

11. When a transformer secondary is short circuited
    a. Secondary current decreases
    b. A small amount of primary current flows
    c. The transformer might be destroyed
    d. Primary current decreases

12. The efficiency of a transformer is
    a. Generally higher than 99.8%
    b. Generally lower than 50%
    c. Generally lower than 70% but higher than 50%
    d. Usually higher than 90%

13. Fig. E-(13) shows a
    a. Step-up autotransformer
    b. Step-down autotransformer
    c. Variac
    d. Potential transformer

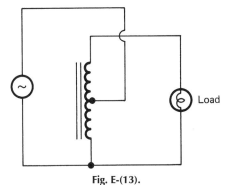

Fig. E-(13).

14. A transformer has a 1200-V primary winding and a 240-V secondary winding. The primary winding has 2000 turns of wire. How many secondary turns does the transformer have?

    a. 400
    b. 10,000
    c. 240
    d. 2000

True or false: Place either T or F in each blank.

_____ 15. The operation of a transformer relies on the principle of mutual inductance.

_____ 16. To increase current capacity of transformers, two transformer windings may be connected in parallel.

_____ 17. The amount of voltage step up or step down of a transformer is determined by the turns ratio of its windings.

_____ 18. The primary winding of a transformer has a voltage rating of 480 V and has 400 turns. The number of turns of wire required on the secondary to produce 120 V is 200.

_____ 19. A 240 V to 24 V step-down transformer draws 1.5 A. The current of the secondary winding is 150 A.

_____ 20. A transformer with N_s = 5000 and N_p = 200 is a *step-down* transformer.

UNIT 5

Frequency-Sensitive AC Circuits

Some types of ac circuits are designed to respond to ac frequencies. Circuits that are used to pass some frequencies and block others are called *frequency-sensitive circuits*. Two types of frequency-sensitive circuits are *filter circuits* and *resonant circuits*. Each type of circuit uses reactive devices to respond to different ac frequencies. These circuits have *frequency response curves*. Frequency is graphed on the horizontal axis and voltage output on the vertical axis. Sample frequency response curves for each type of filter and resonant circuit are shown in the examples that follow. Decibels as they relate to ac circuits are also discussed in this unit.

UNIT OBJECTIVES

Upon completion of this unit, you should be able to:

1. Identify types of filter circuits.
2. Draw response curves for basic filter circuits.
3. Investigate the characteristics of series and parallel resonant circuits.
4. Investigate the frequency response of high-pass, low-pass, and band-pass filters.
5. Verify the operation and characteristics of low-pass, high-pass, and band-pass filters.
6. Define resonance.
7. Calculate resonant frequency.
8. List the characteristics of series and parallel resonant circuits.
9. Define and calculate quality factor (Q) and bandwidth.

Important Terms

Before studying unit 5, you should review the following terms:

Amplification An increase in value.

Attenuation A reduction in value.

Band-pass filter A frequency-sensitive ac circuit that allows incoming frequencies within a certain band to pass through but attenuates frequencies below or above this band.

Bandwidth The band (range) of frequencies that pass easily through a band-pass filter or resonant circuit.

Decibel (dB) A unit used to express an increase or decrease in power, voltage, or current in a circuit.

Filter A circuit used to pass certain frequencies and attenuate all other frequencies.

Frequency The number of ac cycles per second, measured in hertz (Hz).

Frequency response The ability of a circuit to operate over a range of frequencies.

High-pass filter A frequency-sensitive ac circuit that passes high-frequency input signals to its output and attenuates low-frequency signals.

Low-pass filter A frequency-sensitive ac circuit that passes low-frequency input signals to its output and attenuates high-frequency signals.

Parallel resonant circuit A circuit that has an inductor and capacitor connected in parallel to cause response to frequencies applied to the circuit; also called *tank circuit*.

Quality factor (Q) The figure of merit, or ratio of inductive reactance and resistance in a frequency-sensitive circuit.

Resonant circuit See *parallel resonant circuit* and *series resonant circuit*.

Resonant frequency (f_r) The frequency that passes most easily through a frequency-sensitive circuit when $X_L = X_C$ in the circuit. The formula is as follows:

Selectivity The ability of a resonant circuit to select a specific frequency and reject all other frequencies.

Series resonant circuit A circuit that has an inductor and capacitor connected in series to cause response to frequencies applied to the circuit.

Tank circuit See *parallel resonant circuit*.

Filter Circuits

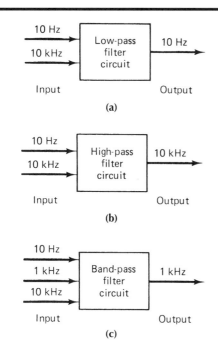

Fig. 5-1. Three types of filter circuits. (a) A low-pass filter passes low frequencies and blocks high frequencies. (b) A high-pass filter passes high frequencies and blocks low frequencies. (c) A band-pass filter passes a midrange of frequencies and blocks high and low frequencies.

The three types of filter circuits are shown in Fig. 5-1. Filter circuits are used to separate one range of frequencies from another. Low-pass filters pass low ac frequencies and block higher frequencies. High-pass filters pass high frequencies and block lower frequencies. Band-pass filters pass a midrange of frequencies and block lower and higher frequencies. All filter circuits have resistance and capacitance *or* inductance.

Figure 5-2 shows the circuits used for low-pass, high-pass, and band-pass filters and their frequency response curves. Many low-pass filters are series *RC* circuits, as shown in Fig. 5-2a. Output voltage (V_{out}) is taken across a capacitor. As frequency increases, capacitive reactance (X_C) decreases, since

$$X_C = \frac{1}{2\pi fC}$$

The voltage drop across the output is equal to *I* times X_C. So as frequency increases, X_C decreases and voltage output decreases.

Series *RL* circuits may be used as low-pass filters. As frequency increases, inductive reactance (X_L) increases because $X_L = 2\pi fL$. Any increase in X_L reduces the current in the circuit. The voltage output taken across the resistor is equal to $I \times R$. So when *I* decreases, V_{out} also decreases. As frequency increases, X_L increases, *I* decreases, and V_{out} decreases.

Figure 5-2b shows two types of high-pass filters. The series *RC* circuit is a common type. The voltage output (V_{out}) is taken across the resistor (R). As frequency increases, X_C decreases. A decrease in X_C causes current flow to increase. The voltage output across the resistor (V_{out}) is equal to $I \times R$. So as *I* increases, V_{out} increases. As frequency increases, X_C decreases, *I* increases, and V_{out} increases.

A series *RL* circuit also may be used as a high-pass filter. V_{out} is taken across the inductor. As frequency increases, X_L increases. V_{out} is equal to *I* times X_L. So as X_L increases, V_{out} also increases. In this circuit, as frequency increases, X_L increases and V_{out} increases.

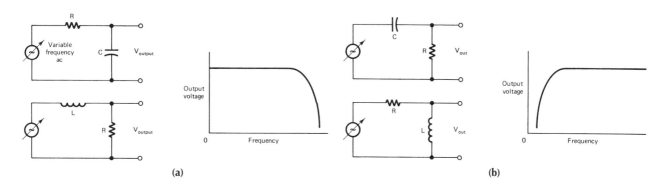

Fig. 5-2. Circuits used to filter ac frequencies and their response curves. (a) Low-pass filters. (b) High-pass filters.

Frequency-sensitive AC Circuits

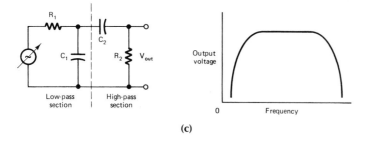

Fig. 5-2. (continued) (c) Band-pass filter.

The band-pass filter of Fig. 5-2c is a combination of low-pass and high-pass filter sections. It is designed to pass a midrange of frequencies and block low and high frequencies. R_1 and C_1 form a low-pass filter and R_2 and C_2 form a high-pass filter. The range of frequencies to be passed is determined by means of calculating the values of resistance and capacitance.

Resonant Circuits

Resonant circuits are designed to pass a range of frequencies and block all others. They have resistance, inductance, and capacitance. Figure 5-3 shows the two types of resonant circuits, series resonant and parallel resonant circuits, and their frequency response curves.

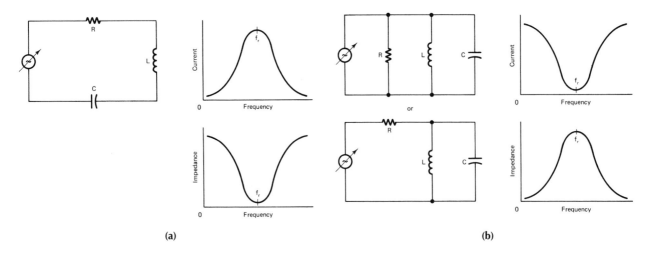

Fig. 5-3. Series (a) and parallel (b) resonant circuits and their frequency response curves.

Series Resonant Circuits

Series resonant circuits are a series arrangement of inductance, capacitance, and resistance. A series resonant circuit offers a small amount of opposition to some ac frequencies and much more opposition to other frequencies. They are important for selecting or rejecting frequencies.

The voltage across inductors and capacitors in ac series circuits are in direct opposition to each other (180° out of phase). They tend to cancel each other out. The frequency applied to a series resonant circuit affects inductive reactance and capacitive reactance. At a specific input frequency, X_L equals X_C. The voltages across the inductor and capacitor are then equal. The total reactive voltage (V_X) is 0 V at this frequency. The opposition offered by the inductor and the capacitor cancel each other at this frequency. The total reactance (X_T) of the circuit ($X_L - X_C$) is zero. The impedance (Z) of the circuit is then equal to the resistance (R).

The frequency at which $X_L = X_C$ is called the *resonant frequency*. To determine the resonant frequency (f_r) of the circuit, use the following formula:

$$f_r = \frac{1}{2\pi\sqrt{LC}}$$

In the formula, L is in henrys, C is in farads, and f_r is in hertz. As either inductance or capacitance increases, resonant frequency decreases. When the resonant frequency is applied to a circuit, a condition called *resonance* exists. Resonance for a series circuit causes the following:

1. $X_L = X_C$
2. $X_T = 0$
3. $V_L = V_C$
4. Total reactive voltage (V_X) equal to zero
5. $Z = R$
6. Total current (I_T) is maximum
7. Phase angle (θ) = 0°

The ratio of reactance (X_L or X_C) to resistance (R) at resonant frequency is called *quality factor* (Q). This ratio is used to determine the range of frequencies or *bandwidth* (BW) a resonant circuit will pass. A sample resonant circuit problem is shown in Fig. 5-4. The frequency range that a resonant circuit will pass (BW) is found by using steps 5 and 6 in Fig. 5-4.

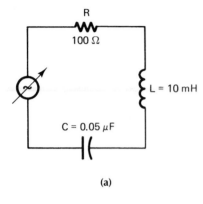

Finding circuit values:

1. Find resonant frequency (f_r):

$$f_r = \frac{1}{2\pi \times \sqrt{L \times C}} = \frac{1}{6.28 \times \sqrt{(10 \times 10^{-3}) \times (0.05 \times 10^{-6})}}$$

$$= \frac{1}{6.28 \times \sqrt{0.5 \times 10^{-9}}} = \frac{1}{6.28 \times (2.23 \times 10^{-5})} = \frac{1}{1.4 \times 10^{-4}} = 7121 \text{ Hz}$$

2. Find X_L^* and X_C at resonant frequency:

$X_L = 2\pi \cdot f \cdot L = 6.28 \times 7121 \times (10 \times 10^{-3}) = 447 \, \Omega$

*Easier to calculate.

3. Find quality factor (Q):

$$Q = \frac{X_L}{R} = \frac{447 \, \Omega}{100 \, \Omega} = 4.47$$

4. Find bandwidth (BW):

$$BW = \frac{f_r}{Q} = \frac{7121 \text{ Hz}}{4.47} = 1593 \text{ Hz}$$

5. Find low-frequency cutoff (f_{lc}):

$f_{lc} = f_r - \frac{1}{2} BW = 7121 \text{ Hz} - 797 \text{ Hz} = 6324 \text{ Hz}$

6. Find high-frequency cutoff (f_{hc})

$f_{hc} = f_r + \frac{1}{2} BW = 7121 + 797 = 7918 \text{ Hz}$

(b)

Fig. 5-4. Sample resonant circuit problem. (a) Circuit. (b) Procedure to find circuit values.

The cutoff points are at about 70% of the maximum output voltage. These are called the low-frequency cutoff (f_{lc}) and high-frequency cutoff (f_{hc}). The bandwidth of a resonant circuit is determined by Q, which is the ratio of X_L and X_C to R. Resistance mainly determines bandwidth. This effect is summarized as follows:

1. When R increases, Q decreases, because $Q = X_L/R$.

2. When Q decreases, BW increases, because $BW = f_r/Q$.

3. When R increases, BW increases.

Two curves in Fig. 5-5 show the effect of resistance on bandwidth. The curve of Fig. 5-5b has high *selectivity*. This means that a resonant circuit with this response curve will select a small range of frequencies. This is very important for radio and television tuning circuits.

Another series resonant circuit problem is shown in Fig. 5-6.

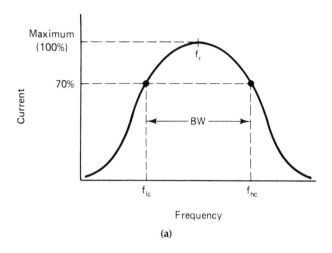

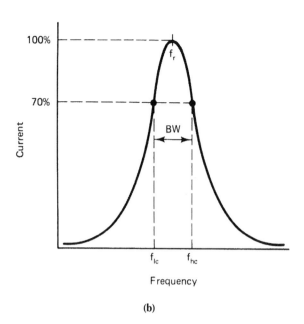

Fig. 5-5. Effect of resistance on bandwidth of a series resonant circuit. (a) High resistance, low selectivity. (b) Low resistance, high selectivity.

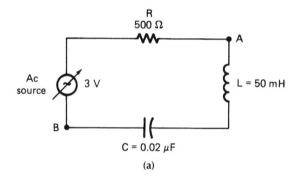

(a)

Fig. 5-6. Series resonant circuit problem.

Finding circuit values:

1. Find resonant frequency (f_r):

$$f_r = \frac{1}{2\pi \times \sqrt{L \times C}} = \frac{1}{6.28 \times \sqrt{(50 \times 10^{-3}) \times (0.02 \times 10^{-6})}}$$

$$= \frac{1}{6.28 \times \sqrt{1.0 \times 10^{-9}}} = \frac{1}{1.9 \times 10^{-4}} = 5035 \text{ Hz}$$

2. Find X_L* and X_C at resonant frequency:
$X_L = 2\pi \cdot f \cdot L = 6.28 \times 5035 \times 0.050 = 1581 \; \Omega$

*Easier to calculate

3. Find quality factor (Q):

$Q = \frac{X_L}{R} = \frac{1581}{500} = 3.162$

4. Find bandwidth (BW):

$BW = \frac{f_r}{Q} = \frac{5035 \text{ Hz}}{3.162} = 1592 \text{ Hz}$

5. Find low-frequency cutoff (f_{LC}):

$f_{lc} = f_r - \frac{1}{2} BW = 5035 - 796 = 4239 \text{ Hz}$

6. Find high-frequency cutoff (f_{hc}):

$f_{hc} = f_r + \frac{1}{2} BW = 5035 + 796 = 5831 \text{ Hz}$

7. Find I_T at resonant frequency:

$I_T = \frac{V_A}{R} = \frac{3 \text{ V}}{500 \; \Omega} = 0.006 \text{ A} = 6 \text{ mA}$

8. Find V_R at resonant frequency:
$V_R = I \times R = 0.006 \text{ A} \times 500 \; \Omega = 3 \text{ V}$

9. Find V_L and V_C at resonant frequency:
$V_L = V_C = I \cdot X_L = 0.006 \text{ A} \times 1581 \; \Omega = 9.49 \text{ V}$*

*Note that this voltage exceeds source voltage—this is called *"voltage magnification."*

(b)

Parallel Resonant Circuits

Parallel resonant circuits are similar to series resonant circuits. Their electric characteristics are somewhat different, but they accomplish the same purpose. Another name for parallel resonant circuits is *tank circuits*. A tank circuit is a parallel combination of L and C used to select or reject ac frequencies.

With the resonant frequency applied to a parallel resonant circuit, the following occurs:

1. $X_L = X_C$
2. $X_T = 0$
3. $I_L = I_C$
4. $I_X = 0$, so the circuit current is minimum
5. $Z = R$ and is maximum
6. Phase angle $(\theta) = 0°$

The calculations used for parallel resonant circuits are similar to those for series circuits. There is one exception. The quality factor (Q) is found with this formula for parallel circuits: $Q = R/X_L$. A parallel resonant circuit problem is shown in Fig. 5-7.

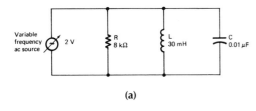

(a)

Finding circuit values:

1. Find resonant frequency (f_r):

$$f_r = \frac{1}{2\pi\sqrt{L \times C}} = \frac{1}{6.28 \times \sqrt{(30 \times 10^{-3}) \times (0.01 \times 10^{-6})}} = \frac{1}{1.0877 \times 10^{-4}} = 9194 \text{ Hz}$$

2. Find X_L and X_C at resonant frequency:

$X_L = 2\pi \cdot f \cdot L = 6.28 \times 9194 \times (30 \times 10^{-3}) = 1732 \; \Omega$

3. Find quality factor (Q):

$$Q = \frac{R}{X_L} = \frac{8000 \; \Omega}{1732 \; \Omega} = 4.62$$

4. Find bandwidth (BW):

$$BW = \frac{f_r}{Q} = \frac{9194 \text{ Hz}}{4.62} = 1990 \text{ Hz}$$

5. Find low cutoff frequency (f_{lc}):

$f_{lc} = f_r - 1/2 \text{ BW} = 9194 \text{ Hz} - 995 \text{ Hz} = 8199 \text{ Hz}$

6. Find high cutoff frequency (f_{hc}):

$f_{hc} = f_r + 1/2 \text{ BW} = 9194 \text{ Hz} + 995 \text{ Hz} = 10{,}189 \text{ Hz}$

7. Find I_R at resonant frequency:

$$I_R = \frac{V_A}{R} = \frac{2 \text{ V}}{8 \text{ k}\Omega} = 0.25 \text{ mA}$$

8. Find I_L and I_C at resonant frequency:

$$I_L = \frac{V_A}{X_L} = \frac{2 \text{ V}}{1732 \; \Omega} = 1.15 \text{ mA}$$

(b)

Fig. 5-7. Parallel resonant circuit problem.

Self-Examination

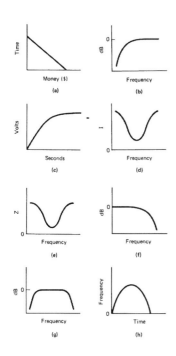

Fig. 5-8.

Match each term with its corresponding diagram in Fig. 5-8. Fill in the letter of the correct diagram.

_____ 1. Low-pass filter

_____ 2. High-pass filter

_____ 3. Band-pass filter

_____ 4. RC time constant curve

_____ 5. Series resonant curve

_____ 6. Parallel resonant curve

7. In a parallel resonant circuit, X_C predominates (is larger than X_L) at frequencies above resonance; X_L predominates at frequencies _____.

8. In a series resonant circuit, X_L predominates at frequencies above resonance; X_C predominates at frequencies _____.

9. The phase angle between current and voltage in a resonant circuit is determined by the predominating influence. When the circuit is inductive, current _____ the _____.

10. Current leads the voltage when the circuit is _____.

11. Two types of resonant circuits are _____ and _____.

12. Three types of filter circuits are _____, _____, and _____.

13. Resonant circuits have _____, _____, and _____.

14. Filter circuits have _____ and _____ or _____.

Answers

1. F	2. B
3. G	4. C
5. D	6. E
7. Below resonance	8. Below resonance
9. Lags voltage	10. Capacitive
11. Series, parallel	12. Low-pass, high-pass, band-pass
13. Resistance, inductance, capacitance	14. Resistance, inductance, capacitance

Decibels

The human ear does not respond to sound levels in the same manner as electronic circuits. An electronic amplifier, for example, has a *linear* rise in signal level. An input signal level of 1 V can produce an output of 10 V. The voltage amplification would be 10:1, or 10. The human ear, however, does not respond in a linear manner. The response is essentially *nonlinear*. As a result of this, sound systems are usually elevated on a *logarithmic* scale. This is an indication of how our ears actually respond to specific signal levels. Gain expressed in logarithms is much more meaningful than linear gain expressed as a relationship.

Logarithms

The logarithm of a given number is the power to which another number, called the base, must be raised to equal the given number. A common logarithm is expressed in terms of powers of 10.

This is illustrated by the following:

$10^4 = 10{,}000$, therefore $\log_{10} 10{,}000 = 4$

$10^3 = 1000$, therefore $\log_{10} 1000 = 3$

$10^2 = 100$, therefore $\log_{10} 100 = 2$

$10^1 = 10$, therefore $\log_{10} 10 = 1$

$10^0 = 1$, therefore $\log_{10} 1 = 0$

This means, for example, that the logarithm of any number between 9999 and 1000 would have a *characteristic* value of 3. The characteristic is an expression of the magnitude range of the number. Numbers between 999 and 100 have a characteristic of 2. Numbers between 99 and 10 have a characteristic of 1. Between 9.0 and 1.0 the characteristic is 0. Number values less than 1.0 have a negative characteristic.

When a number is not an even multiple of 10, its log must have a decimal expression. The decimal part of a logarithm is called the *mantissa*. The logarithm of a number such as 4000 is expressed as 3.6021. The characteristic is 3 because 4000 is between 9999 and 1000. The mantissa of 4000 is 0.6021.

The mantissa is always the same for a given sequence of numbers regardless of the location of the decimal point. For example, the mantissa is the same for 1630, 163.0, 16.3, 1.63, 0.163, and so on. The only difference in the logarithms of these numbers is in the characteristics. The mantissa for 1630 is 0.2122. The logs of the remaining numbers are 3.2122, 2.2122, 1.2122, 0.2122, and 0.2122–1.

Electronic calculators make finding logarithms easy. Simply enter the number into the calculator, and then press the log button. For example, the log of 1590 is 3.2012.

Decibel Applications

The gain of a sound system with several stages of amplification can best be expressed as a ratio of two signal levels. Gain is expressed as the output level divided by the input level. This is determined with the following expression:

$$\text{Power amplification}\ (A_p) = \log\frac{P_{out}}{P_{in}} = \text{bels}$$

For an amplifier with 0.1 W of input and 100 W of output

$$A_p = \log\frac{100\ W}{0.1\ W} = \log 1000 = 3\ \text{bels}$$

The fundamental unit of sound level gain is the *bel* (B). As shown in the calculation, the bel represents a rather large ratio in sound level. A *decibel* (dB) is a more practical measure of sound level. A decibel is one tenth of a bel.

The gain of a single stage of amplification within a system can be determined with decibels. A single amplifier stage can have an input of 10 mW and an output of 150 mW. The power gain would be determined with the following formula:

$$\text{Power amplification}\ (A_p) = 10\ \log\frac{P_{out}}{P_{in}}$$

$$10\ \log\frac{150\ mW}{10\ mW}$$

$$= 10\ \log 15$$

$$= 11.761\ dB$$

The voltage gain of an amplifier can be expressed in decibel values. For this the power level expression must be adapted to accommodate voltage values. The decibel voltage gain formula is as follows:

$$\text{Voltage amplification}\ (A_v) = 20 \times \log\frac{\text{voltage output}}{\text{voltage input}}$$

$$= 20 \log \frac{V_{out}}{V_{in}} \text{(dB)}$$

The logarithm of V_{out}/V_{in} is multiplied by 20 in this equation. Power is expressed as V^2/R. Power gain using voltage and resistance values is expressed as follows:

$$dB = \frac{V^2_{out} \times R_{out}}{V^2_{in} \times R_{in}}$$

If the values of R_{in} and R_{out} are equal, the equation is simplified as follows:

$$dB = \frac{V^2_{out}}{V^2_{in}}$$

The squared voltage values can be expressed as two times the log of the voltage value. Decibel voltage gain therefore becomes the following:

$$dB = 2 \times 10 \log \frac{V_{out}}{V_{in}} \text{ or } 20 \log \frac{V_{out}}{V_{in}}$$

To demonstrate the use of the decibel voltage gain equation, one can apply it to a circuit that has an input voltage of 0.25 V p-p and output voltage of 1.25 V p-p. The voltage gain in decibels is as follows:

$$\text{Voltage gain} = 20 \log \frac{V_{out}}{V_{in}}$$

$$= 20 \log \frac{1.25 \text{ V p-p}}{0.25 \text{ V p-p}}$$

$$= 20 \log 5$$

$$= 20 \times 0.69897$$

$$= 13.9794, \text{ or } 14 \text{ dB}$$

When the decibel value of an amplifier is known, the power gain or voltage gain may be determined by use of *inverse logarithms*, or *antilogarithms*. An antilogarithm is the number from which a logarithm is derived. The process of finding an antilogarithm is the reverse of finding a logarithm. Antilogarithms can be determined easily with a calculator.

As an example of using antilogarithms, assume that an amplifier has a decibel value of +3.5. The power gain of the amplifier is found as follows:

$$dB = 10 \log \frac{P_1}{P_2}$$

$$3.5 = 10 \log \frac{P_1}{P_2}$$

$$0.35 = \log \frac{P_1}{P_2} \text{ (Divide both sides of the equation by 10.)}$$

$$\frac{P_1}{P_2} = \text{antilog } 0.35 \text{ (Find antilogarithm.)}$$

$$\frac{P_1}{P_2} = 2.24$$

The value of 2.24 obtained in the example is the *power ratio*. An amplifier with a gain of +3.5 dB thus has a power gain of 2.24 to 1.

Decibels can be used to express reduction in power or voltage levels. A reduction of input-signal level in a circuit is called *attenuation*. A circuit that attenuates a signal is compared with an amplifier circuit in Fig. 5-9. The decibel value is marked with a (−) sign when the circuit attenuates the input signal. A common example of attenuation occurs in coaxial cable or other signal-transmission cable in which a reduction of signal occurs from input to output.

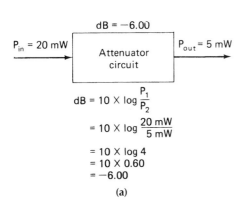

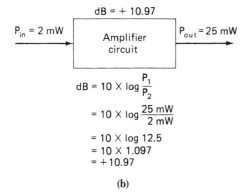

Fig. 5-9. Comparison of attenuator (a) and amplifier circuits (b).

Using Decibels With Filter Circuits

Decibels are commonly used to plot frequency response curves for filter circuits on the type of graph paper shown in Fig. 5-10. One example of a low-pass filter circuit is shown in Fig. 5-11a. The procedure for plotting a frequency response curve for the low-pass circuit is shown in Fig. 5-11b.

Frequency-sensitive AC Circuits 127

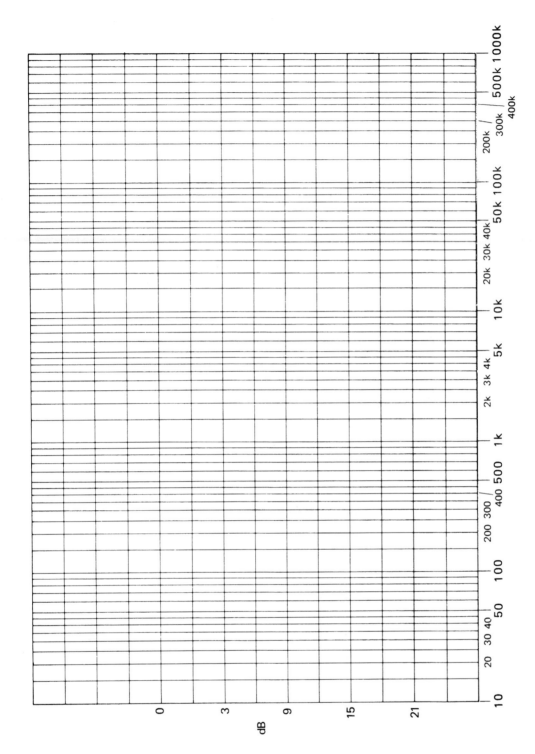

Fig. 5-10. Decibel values used to plot frequency response.

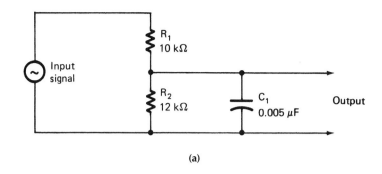

(a)

Finding frequency response:

1. Find the Thévenin equivalent resistance (R_{th}) of R_1 and R_2:

$$R_{th} = R_1 \| R_2 = \frac{10\ k\Omega \times 12\ k\Omega}{10\ k\Omega + 12\ k\Omega} = 5.45\ k\Omega$$

2. Determine the 3-dB frequency using this formula:

$$f_{3dB} = \frac{1}{2\pi \times C \times R_{th}} = \frac{1}{6.28 \times (0.005 \times 10^{-6}) \times (5.45 \times 10^3)}$$

$$= \frac{1}{1.7113 \times 10^{-4}} = 5843\ Hz$$

3. Find the 9-, 15-, and 21-dB frequencies:

$$f_{9dB} = 2 \times f_{3dB} = 2 \times 5843\ Hz = 11{,}686\ Hz$$

$$f_{15dB} = 4 \times f_{3dB} = 4 \times 5843\ Hz = 23{,}372\ Hz$$

$$f_{21dB} = 8 \times f_{3dB} = 8 \times 5843\ Hz = 46{,}744\ Hz$$

4. Label the points on a sheet of frequency-response paper

5. Connect each of the points to form a low-pass frequency-response curve.

(b)

Fig. 5-11. Frequency response for a low-pass filter circuit. (a) Circuit. (b) Calculations.

The selection of decibel values of 3, 9, 15, and 21 dB is standard for plotting frequency response in terms of voltage output of a circuit. These values are easily interpreted by referring to the decibel table in Fig. 5-12. First, locate the 3 dB line. At 3 dB reduction of a signal, the power output is approximately 0.5, or 50%, of the 0 dB reference level, and the voltage is approximately 0.707, or 70.7%, of the 0 dB level. Because the power output of a circuit reduces to about 50% of its original value (0 dB), the 3 dB frequency is called the half-power point.

The selection of decibel points for a high-pass filter circuit is similar to the process used for low-pass filter circuits. A high-pass filter circuit and the procedure for plotting a frequency response curve are shown in Fig. 5-13.

dB	Current or voltage ratio		Power ratio		dB	Current or voltage ratio		Power ratio	
	Gain	Loss	Gain	Loss		Gain	Loss	Gain	Loss
0	1.000	1.0000	1.000	1.0000	3.0	1.413	.7079	1.995	.5012
0.1	1.012	.9886	1.023	.9772	3.1	1.429	.6998	2.042	.4898
0.2	1.023	.9772	1.047	.9550	3.2	1.445	.6918	2.089	.4786
0.3	1.035	.9661	1.072	.9333	3.3	1.462	.6839	2.138	.4677
0.4	1.047	.9550	1.096	.9120	3.4	1.479	.6761	2.188	.4571
0.5	1.059	.9441	1.122	.8913	3.5	1.496	.6683	2.239	.4467
0.6	1.072	.9333	1.148	.8710	3.6	1.514	.6607	2.291	.4365
0.7	1.084	.9226	1.175	.8511	3.7	1.531	.6531	2.344	.4266
0.8	1.096	.9120	1.202	.8318	3.8	1.549	.6457	2.399	.4169
0.9	1.109	.9016	1.230	.8128	3.9	1.567	.6383	2.455	.4074
1.0	1.122	.8913	1.259	.7943	4.0	1.585	.6310	2.512	.3981
1.1	1.135	.8810	1.288	.7762	4.1	1.603	.6237	2.570	.3890
1.2	1.148	.8710	1.318	.7586	4.2	1.622	.6166	2.630	.3802
1.3	1.161	.8610	1.349	.7413	4.3	1.641	.6095	2.692	.3715
1.4	1.175	.8511	1.380	.7244	4.4	1.660	.6026	2.754	.3631
1.5	1.189	.8414	1.413	.7079	4.5	1.679	.5957	2.818	.3548
1.6	1.202	.8318	1.445	.6918	4.6	1.698	.5888	2.884	.3467
1.7	1.216	.8222	1.479	.6761	4.7	1.718	.5821	2.951	.3388
1.8	1.230	.8128	1.514	.6607	4.8	1.738	.5754	3.020	.3311
1.9	1.245	.8035	1.549	.6457	4.9	1.758	.5689	3.090	.3236
2.0	1.259	.7943	1.585	.6310	5.0	1.778	.5623	3.162	.3162
2.1	1.274	.7852	1.622	.6166	5.1	1.799	.5559	3.236	.3090
2.2	1.288	.7762	1.660	.6026	5.2	1.820	.5495	3.311	.3020
2.3	1.303	.7674	1.698	.5888	5.3	1.841	.5433	3.388	.2951
2.4	1.318	.7586	1.738	.5754	5.4	1.862	.5370	3.467	.2884
2.5	1.334	.7499	1.778	.5623	5.5	1.884	.5309	3.548	.2818
2.6	1.349	.7413	1.820	.5495	5.6	1.905	.5248	3.631	.2754
2.7	1.365	.7328	1.862	.5370	5.7	1.928	.5188	3.715	.2692
2.8	1.380	.7244	1.905	.5248	5.8	1.950	.5129	3.802	.2630
2.9	1.396	.7161	1.950	.5129	5.9	1.972	.5070	3.890	.2570

dB	Current or voltage ratio		Power ratio		dB	Current or voltage ratio		Power ratio	
	Gain	Loss	Gain	Loss		Gain	Loss	Gain	Loss
6.0	1.995	.5012	3.981	.2512	10.0	3.162	.3162	10.000	.1000
6.1	2.018	.4955	4.074	.2455	10.1	3.199	.3126	10.23	.09772
6.2	2.042	.4898	4.169	.2399	10.2	3.236	.3090	10.47	.09550
6.3	2.065	.4842	4.266	.2344	10.3	3.273	.3055	10.72	.09333
6.4	2.089	.4786	4.365	.2291	10.4	3.311	.3020	10.96	.09120
6.5	2.113	.4732	4.467	.2239	10.5	3.350	.2985	11.22	.08913
6.6	2.138	.4677	4.571	.2188	10.6	3.388	.2951	11.48	.08710
6.7	2.163	.4624	4.677	.2138	10.7	3.428	.2917	11.75	.08511
6.8	2.188	.4571	4.786	.2089	10.8	3.467	.2884	12.02	.08318
6.9	2.213	.4519	4.898	.2042	10.9	3.508	.2851	12.30	.08128
7.0	2.239	.4467	5.012	.1995	11.0	3.548	.2818	12.59	.07943
7.1	2.265	.4416	5.129	.1950	11.1	3.589	.2786	12.88	.07762
7.2	2.291	.4365	5.248	.1905	11.2	3.631	.2754	13.18	.07586
7.3	2.317	.4315	5.370	.1862	11.3	3.673	.2723	13.49	.07413
7.4	2.344	.4266	5.495	.1820	11.4	3.715	.2692	13.80	.07244

Fig. 5-12. Decibel table.

dB	Current or voltage ratio		Power ratio		dB	Current or voltage ratio		Power ratio	
	Gain	Loss	Gain	Loss		Gain	Loss	Gain	Loss
7.5	2.371	.4217	5.623	.1778	11.5	3.758	.2661	14.13	.07079
7.6	2.399	.4169	5.754	.1738	11.6	3.802	.2630	14.45	.06918
7.7	2.427	.4121	5.888	.1698	11.7	3.846	.2600	14.79	.06761
7.8	2.455	.4074	6.026	.1660	11.8	3.890	.2570	15.14	.06607
7.9	2.483	.4027	6.166	.1622	11.9	3.936	.2541	15.49	.06457
8.0	2.512	.3981	6.310	.1585	12.0	3.981	.2512	15.85	.06310
8.1	2.541	.3936	6.457	.1549	12.1	4.027	.2483	16.22	.06166
8.2	2.570	.3890	6.607	.1514	12.2	4.074	.2455	16.60	.06026
8.3	2.600	.3846	6.761	.1479	12.3	4.121	.2427	16.98	.05888
8.4	2.630	.3802	6.918	.1445	12.4	4.169	.2399	17.38	.05754
8.5	2.661	.3758	7.079	.1413	12.5	4.217	.2371	17.78	.05623
8.6	2.692	.3715	7.244	.1380	12.6	4.266	.2344	18.20	.05495
8.7	2.723	.3673	7.413	.1349	12.7	4.315	.2317	18.62	.05370
8.8	2.754	.3631	7.586	.1318	12.8	4.365	.2291	19.05	.05248
8.9	2.786	.3589	7.762	.1288	12.9	4.416	.2265	19.50	.05129
9.0	2.818	.3548	7.943	.1259	13.0	4.467	.2239	19.95	.05012
9.1	2.851	.3508	8.128	.1230	13.1	4.519	.2213	20.42	.04898
9.2	2.884	.3467	8.318	.1202	13.2	4.571	.2188	20.89	.04786
9.3	2.917	.3428	8.511	.1175	13.3	4.624	.2163	21.38	.04677
9.4	2.951	.3388	8.710	.1148	13.4	4.677	.2138	21.88	.04571
9.5	2.985	.3350	8.913	.1122	13.5	4.732	.2113	22.39	.04467
9.6	3.020	.3311	9.120	.1096	13.6	4.786	.2089	22.91	.04365
9.7	3.055	.3273	9.333	.1072	13.7	4.842	.2065	23.44	.04266
9.8	3.090	.3236	9.550	.1047	13.8	4.898	.2042	23.99	.04169
9.9	3.126	.3199	9.772	.1023	13.9	4.955	.2018	24.55	.04074

dB	Current or voltage ratio		Power ratio		dB	Current or voltage ratio		Power ratio	
	Gain	Loss	Gain	Loss		Gain	Loss	Gain	Loss
14.0	5.012	.1995	25.12	.03981	17.5	7.499	.1334	56.23	.01778
14.1	5.070	.1972	25.70	.03890	17.6	7.586	.1318	57.54	.01738
14.2	5.129	.1950	26.30	.03802	17.7	7.674	.1303	58.88	.01698
14.3	5.188	.1928	26.92	.03715	17.8	7.762	.1288	60.26	.01660
14.4	5.248	.1905	27.54	.03631	17.9	7.852	.1274	61.66	.01622
14.5	5.309	.1884	28.18	.03548					
14.6	5.370	.1862	28.84	.03467	18.0	7.943	.1259	63.10	.01585
14.7	5.433	.1841	29.51	.03388	18.1	8.035	.1245	64.57	.01549
14.8	5.495	.1820	30.20	.03311	18.2	8.128	.1230	66.07	.01514
14.9	5.559	.1799	30.90	.03236	18.3	8.222	.1216	67.61	.01479
					18.4	8.318	.1202	69.18	.01445
15.0	5.623	.1778	31.62	.03162	18.5	8.414	.1189	70.79	.01413
15.1	5.689	.1758	32.36	.03090	18.6	8.511	.1175	72.44	.01380
15.2	5.754	.1738	33.11	.03020	18.7	8.610	.1161	74.13	.01349
15.3	5.821	.1718	33.88	.02951	18.8	8.710	.1148	75.86	.01318
15.4	5.888	.1698	34.67	.02884	18.9	8.811	.1135	77.62	.01288
15.5	5.957	.1679	35.48	.02818					
15.6	6.026	.1660	36.31	.02754	19.0	8.913	.1122	79.43	.01259
15.7	6.095	.1641	37.15	.02692	19.1	9.016	.1109	81.28	.01230
15.8	6.166	.1622	38.02	.02630	19.2	9.120	.1096	83.18	.01202
15.9	6.237	.1603	38.90	.02570	19.3	9.226	.1084	85.11	.01175

Fig. 5-12. Decibel table. (continued)

dB	Current or voltage ratio		Power ratio		dB	Current or voltage ratio		Power ratio	
	Gain	Loss	Gain	Loss		Gain	Loss	Gain	Loss
16.0	6.310	.1585	39.81	.02512	19.4	9.333	.1072	87.10	.01148
16.1	6.383	.1567	40.74	.02455	19.5	9.441	.1059	89.13	.01122
16.2	6.457	.1549	41.69	.02399	19.6	9.550	.1047	91.20	.01096
16.3	6.531	.1531	42.66	.02344	19.7	9.661	.1035	93.33	.01072
16.4	6.607	.1514	43.65	.02291	19.8	9.772	.1023	95.50	.01047
16.5	6.683	.1496	44.67	.02239	19.9	9.886	.1012	97.72	.01023
16.6	6.761	.1479	45.71	.02188	20.0	10.00	.1000	100.00	.01000
16.7	6.839	.1462	46.77	.02138	30.0	31.62	.0316	10^3	10^{-3}
16.8	6.918	.1445	47.86	.02089	40.0	100.00	.0100	10^4	10^{-4}
16.9	6.998	.1429	48.98	.02042	50.0	316.2	.0032	10^5	10^{-5}
					60.0	10^3	10^{-3}	10^6	10^{-6}
17.0	7.079	.1413	50.12	.01995	80.0	10^4	10^{-4}	10^8	10^{-8}
17.1	7.161	.1396	51.29	.01950	100.0	10^5	10^{-5}	10^{10}	10^{-10}
17.2	7.244	.1380	52.48	.01905	120.0	10^6	10^{-6}	10^{12}	10^{-12}
17.3	7.328	.1365	53.70	.01862	140.0	10^7	10^{-7}	10^{14}	10^{-14}
17.4	7.413	.1349	54.95	.01820	180.0	10^9	10^{-9}	10^{18}	10^{-18}

Fig. 5-12. Decibel table. (continued)

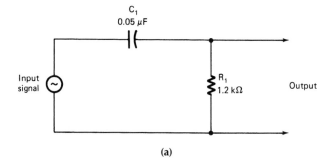

(a)

Finding frequency response:

1. Determine the 3-dB frequency:

$$f_{3dB} = \frac{1}{2\pi \times R \times C} = \frac{1}{6.28 \times (1.2 \times 10^3) \times (0.05 \times 10^{-6})}$$

$$= \frac{1}{3.768 \times 10^{-4}} = 2654 \text{ Hz}$$

2. Find the 9-, 15-, and 21-dB frequencies:

$f_{9dB} = f_{3dB} \div 2 = 2654 \div 2 = 1327$ Hz

$f_{15dB} = f_{3dB} \div 4 = 2654 \div 4 = 663.5$ Hz

$f_{21dB} = f_{3dB} \div 8 = 2654 \div 8 = 332$ Hz

3. Label each of the points on a sheet of frequency-response paper.

4. Connect each of the points to form a high-pass frequency-response curve.

(b)

Fig. 5-13. Frequency response for a high-pass filter circuit.
(a) Circuit. (b) Calculations.

Band-pass filter circuits are a combination of low-pass and high-pass filter circuits. An example of a band-pass filter circuit and the procedure for plotting a frequency response curve are shown in Fig. 5-14. Notice the use of the combined resistances ($R_1 + R_2$) for the high-pass section and the use of the Thevenin equivalent resistance (R_{TH}) for the low-pass section. The 3 dB frequency on the low-frequency end of the response curve is called the *low-cutoff frequency* (f_{lc}). The high-frequency 3 dB point is called the *high-cutoff frequency* (f_{hc}).

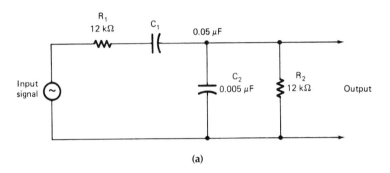

(a)

Finding frequency response:

1. Determine the 3-, 9-, 15-, and 21-dB frequencies for the high-pass section:

$$f_{3dB} = \frac{1}{2\pi \times R \times C} = \frac{1}{6.28 \times (24 \times 10^3) \times (0.05 \times 10^{-6})}$$

$$= \frac{1}{7.536 \times 10^{-3}} = 133 \text{ Hz } (f_{lc})$$

$f_{9dB} = f_{3dB} \div 2 = 66.5$ Hz

$f_{15dB} = f_{3dB} \div 4 = 33.25$ Hz

$f_{21dB} = f_{3dB} \div 8 = 16.6$ Hz

2. Determine the 3-, 9-, 15-, and 21-dB frequencies for the low-pass section:

$$f_{3dB} = \frac{1}{2\pi \times R_{th} \times C} = \frac{1}{6.28 \times (6 \times 10^3) \times (0.005 \times 10^{-6})}$$

$$= \frac{1}{1.884 \times 10^{-4}} = 5308 \text{ Hz } (f_{hc})$$

$f_{9dB} = f_{3dB} \times 2 = 10,616$ Hz

$f_{15dB} = f_{3dB} \times 4 = 21,232$ Hz

$f_{21dB} = f_{3dB} \times 8 = 42,464$ Hz

3. Label each of the points on a sheet of frequency-response paper.

4. Connect each of the points to form a band-pass frequency-response curve.

(b)

Fig. 5-14. Frequency response for a band-pass filter circuit.
(a) Circuit. (b) Calculations.

Waveshaping Control

A circuit designed to change the shape of a waveform is very essential in electronics today. The type of waveshape achieved is based primarily on the shape of the applied signal and the characteristics of the shaping circuit. Sine, square, and sawtooth waves are the basic input waves to be processed. Square and sawtooth waveforms are shown in Fig. 5-15. Waves may be of the periodic type or have a pulsing characteristic. A *periodic wave* continually changes value within a given time frame. *Pulse waveforms* occur momentarily at intervals. *Repetitive pulses* occur at definite time periods. *Transient pulses* occur randomly and are the result of a switching action or some other electric change. Waveshaping circuits may be used to minimize transient pulse conditions so that they will not cause irregular control of a circuit.

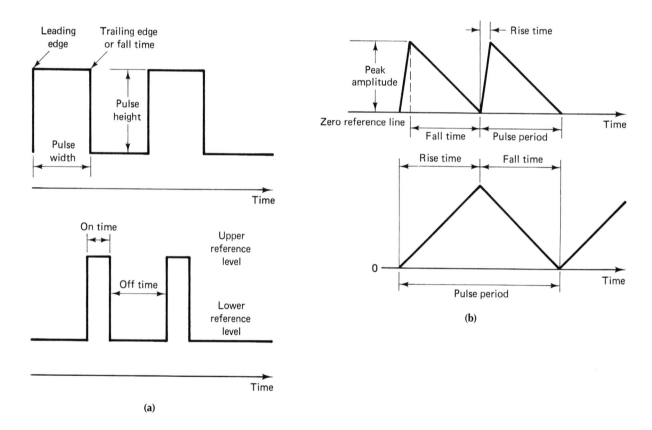

Fig. 5-15. Square or rectangular waveforms (a) and sawtooth waveforms (b).

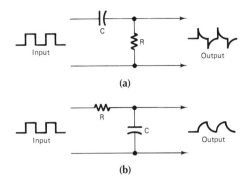

Fig. 5-16. *RC* waveshaping circuits. (a) Differentiator circuit. (b) Integrator circuit.

Numerous devices and components have been developed and are being used to achieve various waveshaping operations. Waveshaping is considered to be an altering process that changes the basic form of a wave or its conduction time or causes a variation in its amplitude.

A rather simple method of altering the basic form of a wave is to apply it to an *RC* time constant circuit. Two distinct changes can be produced according to the *RC* component combination selected. With a square or rectangular wave applied, a differentiated or integrated waveshape may be produced (Fig. 5-16).

Differentiator Circuits

The *differentiator* circuit in Fig. 5-16a has a square-wave input signal applied to a capacitor. The output is developed across a resistor. Any abrupt change in the input signal causes a pronounced change or difference in the output voltage developed across the resistor. The term *differentiator* is used to describe the changing output of this circuit.

When the leading edge of an applied square wave rises quickly, it causes the capacitor of a differentiator to charge immediately. The series-connected resistor of this circuit "sees" a current identical to that produced by the capacitor. This current rises sharply to a peak and then drops off very quickly when the capacitor becomes charged. Current passing through the resistor develops an output voltage that conforms to the charging current of the capacitor.

When the trailing edge of an applied square wave is reached, the voltage applied to the capacitor drops abruptly. The electrostatic charge on the capacitor begins to discharge back through the resistor. This current is in a direction opposite that of the original charging current. The resulting current through the resistor causes a corresponding output voltage to be developed. The polarity of this voltage is opposite that of the charging cycle. The output voltage drops below the zero reference line.

When a differentiator has a short time constant with respect to the applied square wave, the output has only sharp positive and negative spikes. A longer time constant, however, causes the trailing edge of the output to be extended somewhat. This indicates that the discharge action of the capacitor has been prolonged. The *RC* time constant of a differentiator with respect to the pulse repetition rate of the square-wave input largely determines the form of the output of this circuit. The reciprocal of the time that it takes for a wave of this type to repeat itself is called the *pulse repetition rate* (PRR). This term compares with the frequency of a sine wave, which is expressed in hertz.

Integrator Circuits

The *integrator* in Fig. 5-16b is a waveshaping circuit that has the input signal applied to a resistor and the output developed across a capacitor. The value of the resistor in this case tends to restrict the charging time of the capacitor. The voltage developed across the capacitor therefore has a gradual rise time and fall time when a square wave is applied.

The output of an integrator can be altered by changing the PRR of the input or changing the time constant of the RC network. Time constants longer than the on time of a square wave tend to cause the wave to have a sawtooth appearance. A shorter time constant tends to cause a slope on the leading and trailing edges followed by straight-line areas. Integrators are used primarily to sum, or total, signal values that are variable.

Self-Examination

Filter Circuits

Solve the following problems, which deal with filter circuits.

15. Refer to Fig. 5-2a. Use values of $R_1 = 20$ kΩ, $R_2 = 15$ kΩ, and $C_1 = 0.004$ µF for a low-pass filter circuit. Determine the 3 dB, 9 dB, 15 dB, and 21 dB frequencies. Plot a frequency response curve.

16. Refer to Fig. 5-2b. Use values of $R_1 = 2$ kΩ and $C_1 = 0.03$ µF for a high-pass filter circuit. Determine the 3 dB, 9 dB, 15 dB, and 21 dB frequencies. Plot a frequency response curve.

17. Refer to Fig. 5-2c. Use values of $R_1 = 10$ kΩ, $R_2 = 6$ kΩ, $C_1 = 0.04$ µF, and $C_2 = 0.0025$ µF for a band-pass filter circuit. Determine the low and high 3 dB, 9 dB, 15 dB, and 21 dB frequencies. Plot a frequency response curve.

Resonant Circuit

Solve the following problems, which deal with resonant circuits.

18. Refer to the series resonant circuit of Fig. 5-6. Use values of $R = 1$ kΩ, $L = 100$ mH, $V_{in} = 2$ V, and $C = 0.025$ µF. Determine the following: (a) resonant frequency, (b) X_L at resonant frequency, (c) quality factor, (d) bandwidth, (e) low-frequency cutoff, (f) high-frequency cutoff, (g) I_T at resonant frequency, (h) V_R at f_r, and (i) V_L at f_r.

19. Refer to the parallel resonant circuit of Fig. 5-7. Use values of $R = 5$ kΩ, $l = 20$ mH, $C = 0.02$ µF, and $V_{in} = 1$ V. Determine the following: (a) f_r, (b) X_L at f_r, (c) Q, (d) BW, (e) f_{lc}, (f) f_{hc}, (g) I_R at f_r, and (h) I_L at f_r.

Decibel Problems

Solve the following problems, which deal with decibels.

20. A circuit has a loss of −16 dB. What power ratio corresponds to this loss?

21. The input to a power line is 220 mW. The power delivered at the end of the line is 40 mW. What is the attenuation in decibels?

22. A power input of an amplifier is 20 mW. The output is 120 mW. What is the amplification in decibels?

23. What is the ratio of output power to input power of a circuit if there is a gain of 15 dB?

24. A coaxial transmission line has a power loss of −22 dB. Determine the power ratio.

25. A circuit has a loss of −12 dB. What voltage ratio corresponds to this loss?

26. The input to a circuit is 180 mV. The output voltage is 30 mV. What is the decibel attenuation?

27. The voltage input to an amplifier circuit is 15 mV. The output is 120 mV. What is the amplification in decibels?

28. What is the voltage ratio of a circuit if there is a gain of 8 dB?

Answers

15. 3 dB = 3791 Hz
    9 dB = 7582 Hz
    15 dB = 15,164 Hz
    21 dB = 30,328 Hz

16. 3 dB = 2654 Hz
    9 dB = 1327 Hz
    15 dB = 663.5 Hz
    21 dB = 331.75 Hz

17. 3 dB = 1061 and 10,616 Hz
    9 dB = 530 and 21,231 Hz
    15 dB = 265 and 42,463 Hz
    21 dB = 132 and 84,926 Hz

18. a. f_r = 3185 Hz
    b. X_L = 2000 Ω
    c. Q = 2
    d. BW = 1592.5 Hz
    e. f_{lc} = 2388.75 Hz
    f. f_{hc} = 3981.25 Hz
    g. I_T = 0.001 A = 1 mA
    h. V_R = 1 V
    i. V_L = 2 V

19. a. f_r = 7962 Hz
    b. X_L = 1000 Ω
    c. Q = 5
    d. BW = 1592 Hz
    e. f_{lc} = 7166 Hz
    f. f_{hc} = 8758 Hz
    g. I_p = 0.2 mA
    h. I_L = 1 mA

20. 39.81

21. −7.4 dB
22. +7.78 dB
23. 31.62
24. 158.49
25. 3.98
26. −15.56 dB
27. +18.06 dB
28. 2.51

EXPERIMENT 5-1

LOW-PASS FILTER CIRCUITS

The purpose of this experiment is to analyze the theoretical and experimental operation of low-pass filters.

OBJECTIVES

1. To calculate a frequency response curve for a low-pass filter circuit.

2. To make measurements with an oscilloscope or multimeter for plotting an experimental frequency response curve.

3. To use a signal generator or function generator as a voltage source.

EQUIPMENT

VOM (multimeter)

Audio signal generator or function generator

Capacitor: 0.005 µF

Resistors: 10 kΩ, 15 kΩ

Connecting wires

PROCEDURE

1. Construct the circuit shown in Fig. 5-1A.

2. The following steps are to be used in the preparation of a theoretical frequency response curve:

    a. Theveninize the circuit at the break points shown in Fig. 5-1A. Draw the Thevenin equivalent circuit (see *Understanding DC Circuits*).

    b. To determine the 3 dB frequency, use the formula: $f_{3dB} = 1/2\pi C R_{TH}$, where f is in hertz, C is in farads, and R_{TH} is in ohms. f_{3dB} = Hz.

    c. Label the axes of a piece of semilog graph paper as shown in Fig. 5-1B. Using your data, locate and identify the point (f_{3dB}, 3 dB) on the graph (Fig. 5-1C). Locate and identify the point ($2 \times f_{3dB}$, 9 dB). Locate and identify the point ($4 \times f_{3dB}$, 15 dB). Locate and identify the point ($8 \times f_{3dB}$, 21 dB). Connect the points to form a curve. This is the theoretical frequency response curve of the circuit of Fig. 5-1A.

3. Set the signal generator frequency at 100 Hz. Maintain the generator output level at 2 V throughout the experiment.

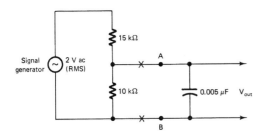

Fig. 5-1A.

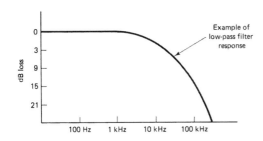

Fig. 5-1B.

Frequency-sensitive AC Circuits

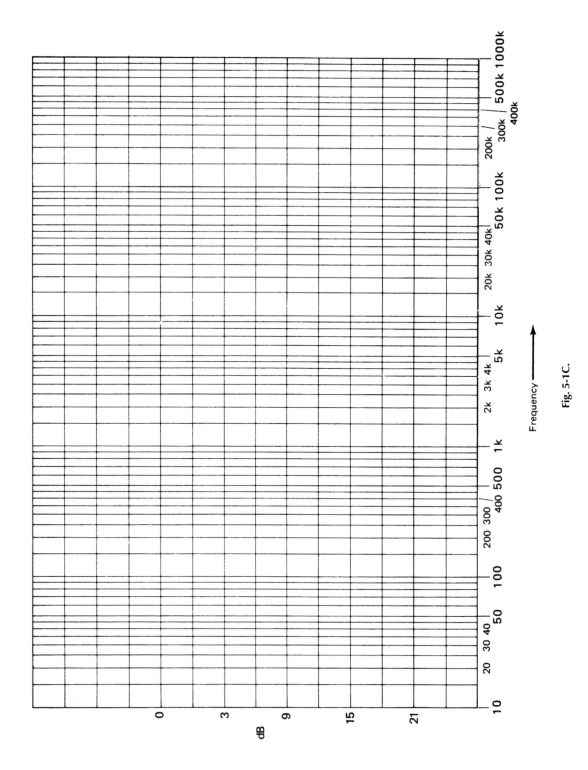

Fig. 5-1C.

4. Place an oscilloscope across the capacitor and adjust the display for a suitable height. Increase the generator frequency until the scope display has been reduced to 70.7% of the original height. This is the 3 dB point. Record the frequency at which it occurred. Develop a curve that represents the experimental frequency response of the circuit. Plot this curve on the same graph paper. Values of frequency should range from 100 Hz to approximately 20 kHz.

5. Compare the calculated and measured 3 dB frequencies.

6. When the generator is set to twice the 3 dB frequency, does the resulting frequency coincide with the theoretical 9 dB point on the theoretical curve? _____

7. If the generator frequency is again doubled, does this frequency coincide with the 15 dB point of the curve? _____

8. Again double the generator frequency and determine whether it coincides with the 21 dB point on the curve. _____

9. Complete the frequency response curves on the semilog graph of Fig. 5-1C showing the theoretical and experimental curves for the low-pass filter you constructed.

ANALYSIS

1. Discuss the operation of a low-pass filter.
   _____
   _____
   _____
   _____
   _____
   _____
   _____
   _____
   _____
   _____
   _____
   _____
   _____

EXPERIMENT 5-2

HIGH-PASS FILTER CIRCUITS

In this experiment, you will observe the characteristics of a high-pass filter circuit and develop a theoretical and experimental frequency response curve.

OBJECTIVES

1. To calculate a theoretical frequency response curve for a high-pass filter circuit.

2. To make measurements with an oscilloscope or multimeter for plotting an experimental frequency response curve.

EQUIPMENT

VOM (multimeter)

Audio signal generator or function generator

Capacitor: 0.05 µF

Resistor: 1.5 kΩ

Connecting wires

PROCEDURE

1. Refer to Fig. 5-2A. Use the same technique as in experiment 5-1 to construct a theoretical high-pass frequency response curve, except use the following expression to determine the 3 dB point:

$$f_{3dB} = \frac{1}{2\pi RC} \qquad R = 1.5 \text{ k}\Omega$$

The additional points to be identified are (f_{3dB}, 3 dB), ($f_{3dB}/2$, 9 dB), ($f_{3dB}/4$, 15 dB), and ($f_{3dB}/8$, 21 dB).

2. Construct the circuit shown in Fig. 5-2A and set the signal generator frequency to 25 kHz. Maintain the generator output at 2 V RMS.

3. Place a scope across the resistor and adjust the display for a convenient height. Decrease the frequency until the scope display has been reduced to 70.7% of the original height. This is the actual 3 dB point. Record this value. Using Fig. 5-2B (semilog paper), make a frequency response plot of the circuit. Values should range from about 100 Hz to 25 kHz.

4. Compare the calculated and the measured 3 dB frequencies. _____

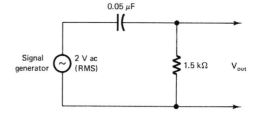

Fig. 5-2A.

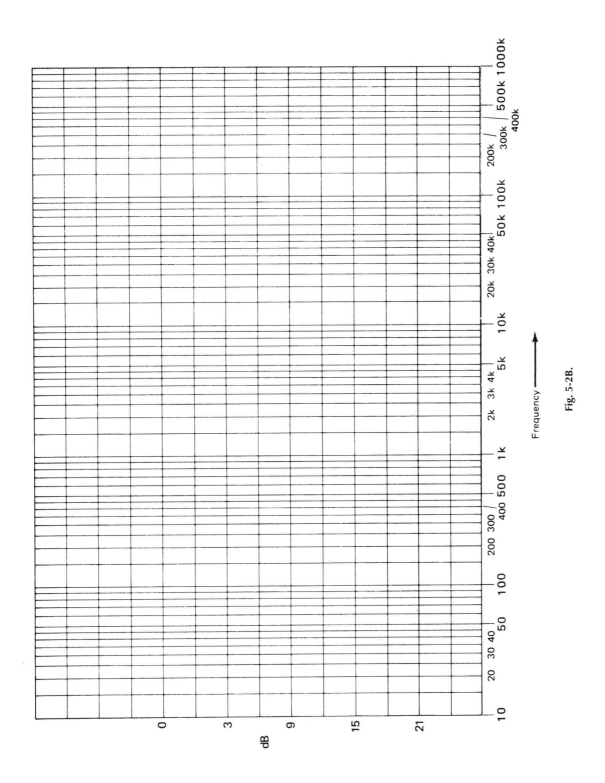

Fig. 5-2B.

5. Record the measured values of the 9 dB, 15 dB, and 21 dB points on the graph.

6. Plot curves on the semilog paper of Fig. 5-2B showing the theoretical and experimental values of frequency response for this high-pass filter circuit.

ANALYSIS

1. Discuss the operation of a high-pass filter circuit.

EXPERIMENT 5-3

BAND-PASS FILTER CIRCUITS

In this experiment, you will observe the effect of using a combination high-pass and low-pass filters to form what is called a band-pass filter circuit. Such circuits are commonly encountered in radios, televisions, and other communication equipment.

OBJECTIVES

1. To calculate a theoretical frequency response curve for a band-pass filter circuit.

2. To make measurements with an oscilloscope or multimeter for plotting an experimental frequency response curve for a band-pass filter circuit.

EQUIPMENT

VOM (multimeter)

Audio signal generator or function generator

Capacitors: 0.05 µF, 0.005 µF

Resistors: 10 kΩ, 15 kΩ

Connecting wires

PROCEDURE

1. To demonstrate a band-pass condition, construct the circuit shown at the top of Fig. 5-3A.

2. Calculate the theoretical high and low 3 dB points by analyzing the circuit at a low frequency and then at a high frequency. Refer to the procedure of Fig. 5-3B.

3. Initially set the signal generator to 2 kHz and maintain its output level at 2 V throughout the experiment.

4. Place a scope across the output and vary the generator frequency until the low and high 3 dB points are found. The frequency range of the band pass is _____ to _____.

5. Make a frequency response curve from the experimental data obtained. Also prepare a theoretical frequency response curve on the graph of Fig. 5-3C.

ANALYSIS

1. Discuss the operation of a band-pass filter circuit. _____
_____
_____
_____
_____
_____

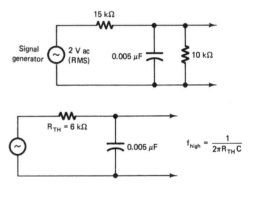

Fig. 5-3A.

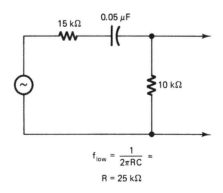

Fig. 5-3B.

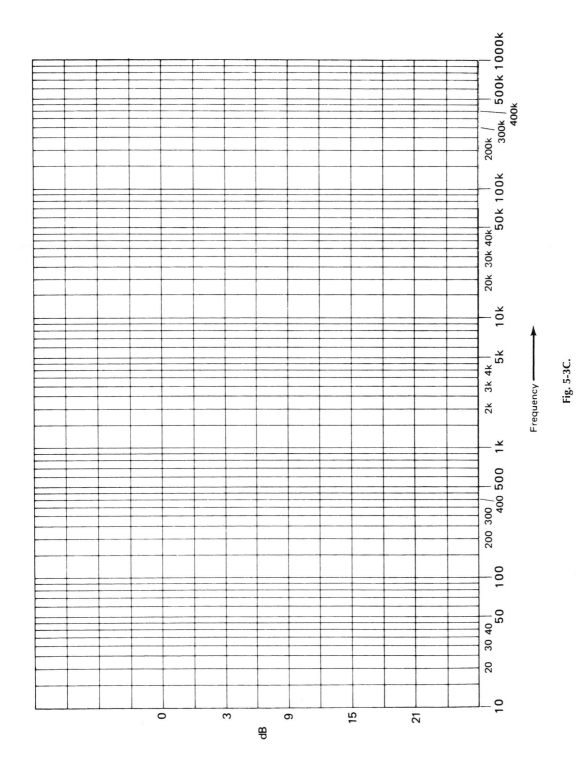

Fig. 5-3C.

EXPERIMENT 5-4

SERIES RESONANT CIRCUITS

A series resonant circuit is a frequency-sensitive circuit. It consists of a series arrangement of inductance, capacitance, and resistance. The series resonant circuit is frequency-sensitive because it offers a small amount of opposition to certain frequencies of alternating current and much greater opposition to other frequencies. Such circuits are important for selecting or rejecting specific frequencies or ranges of frequencies.

OBJECTIVE

To study the characteristics of a series resonant circuit and to observe the quality factor (Q) and bandwidth of this type of circuit.

EQUIPMENT

> Signal generator
>
> Inductor: 107 mH air core
>
> Capacitors: 0.1 µF, 0.47 µF
>
> Resistor: 470 Ω
>
> Oscilloscope
>
> VOM (multimeter)

PROCEDURE

1. Construct the series resonant circuit shown in Fig. 5-4A.

2. Calculate the resonant frequency for this circuit.

3. Turn on the signal generator and adjust the frequency to approximately the calculated resonant frequency. Keep the amplitude adjusted to 1 V RMS.

4. Place the vertical input probes of an oscilloscope across L and C (points B and C). Adjust the horizontal gain of the oscilloscope so that a suitable trace is shown on the screen.

5. Vary the frequency of the signal generator until the amplitude of the vertical line shown on the oscilloscope is minimum. The minimum amplitude corresponds to the measured resonant frequency: measured resonant frequency f_r = _____ Hz.

6. Turn off the signal generator and replace the 0.1 µF capacitor with a 0.47 µF capacitor.

7. Calculate the new resonant frequency.

8. Measure the resonant frequency using the procedure of steps 4 and 5. Measured resonant frequency f_r = _____ Hz.

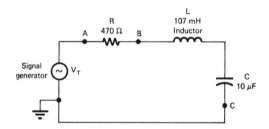

Fig. 5-4A. Series *RLC* circuit.

9. Place the oscilloscope probes across points A and B.

10. Vary the signal generator above and below the resonant frequency while observing the waveform on the oscilloscope. What occurs? _____

11. Adjust the signal generator to the resonant frequency.

12. Remove the oscilloscope from the circuit.

13. Measure the following values with the oscilloscope properly calibrated or with a VOM:
    $V_R =$ _____ V ac; $V_L =$ _____ V ac;
    $V_C =$ _____ V ac.

14. Adjust the signal generator to a frequency approximately 40% above the resonant frequency.

15. Measure the following values:
    $V_R =$ _____ V ac; $V_L =$ _____ V ac;
    $V_C =$ _____ V ac.

16. Adjust the signal generator to a frequency approximately 40% below the resonant frequency.

17. Measure the voltages once again.
    $V_R =$ _____ V ac; $V_L =$ _____ V ac;
    $V_C =$ _____ V ac.

ANALYSIS

1. Compare the calculated and measured values of resonant frequency from steps 2 and 5.

2. What are some reasons for the difference between measured and calculated resonant frequency? _____

3. How would you use a voltmeter to determine resonant frequency of a circuit? _____

4. Calculate the quality factor (Q) of the original circuit.

5. Calculate the bandwidth of the original circuit.

6. How do the values of V_R, V_L, and V_C vary as the frequency is moved to values above and below resonance (see steps 12–17)? V_R: _____;
   V_L: _____; V_C: _____.

7. Complete the following statements: In a series resonant circuit with the resonant frequency applied to the circuit, impedance is _____, current is _____, phase angle is _____, X_L = _____, and R = _____.

8. What effect does increasing resistance have on a series resonant circuit? _____

EXPERIMENT 5-5

PARALLEL RESONANT CIRCUITS

A parallel resonant circuit is another type of frequency-sensitive circuit. This type of circuit is similar in function to a series resonant circuit; however, its electric characteristics are very different. Another name used to describe the parallel resonant circuit configuration is *tank circuit*. The tank circuit is a parallel combination of L and C that may be used to select or reject specific frequencies of alternating current.

OBJECTIVE

To study the characteristics of parallel resonant circuits when variable frequencies of alternating current are applied.

EQUIPMENT

Signal generator

Capacitors: 0.01 µF, 0.1 µF

Inductor: 107 mH air core

Resistor: 1 kΩ

Oscilloscope

VOM (multimeter)

PROCEDURE

1. Construct the parallel resonant circuit shown in Fig. 5-5A.

2. Calculate the resonant frequency for this circuit.

$$f_r = \frac{1}{2\pi\sqrt{LC}} = \underline{\qquad} \text{ Hz}$$

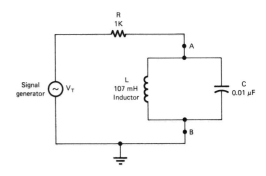

Fig. 5-5A. Parallel *RLC* resonant circuit.

3. Turn on the oscilloscope and connect the vertical input probes across L and C (points A to B).

4. Adjust the signal generator to an amplitude of 1 V RMS at the calculated resonant frequency.

5. Adjust the signal generator until the display shown on the oscilloscope is maximum in amplitude. This value corresponds to the measured resonant frequency. Measured resonant frequency f_r = _____ Hz.

6. Turn off the signal generator and exchange the 0.01 µF capacitor for a 0.1 µF capacitor.

7. Calculate the new resonant frequency

$$f_r = \frac{1}{2\pi\sqrt{LC}} = \underline{\qquad} \text{ Hz}$$

8. Measure the resonant frequency using the procedure of steps 4 and 5. Measured resonant frequency f_r = _____ Hz.

9. Use an ac milliammeter to measure the following values of ac current with the signal generator adjusted to the resonant frequency:
$I_R =$ _____ mA; $I_L =$ _____ mA;
$I_C =$ _____ mA. Or calculate as:
$I_R = V_R/R =$ _____ mA;
$I_L = V_L/X_L =$ _____ mA;
$I_C = V_C/X_C =$ _____ mA.

10. Adjust the signal generator to a frequency approximately 40% above the resonant frequency. Measure the following values:
$I_R =$ _____ mA; $I_L =$ _____ mA;
$I_C =$ _____ mA. Or calculate as:
$I_R = V_R/R =$ _____ mA;
$I_L = V_L/X_L =$ _____ mA;
$I_C = V_C/X_C =$ _____ mA.

11. Adjust the signal generator to a frequency approximately 40% below the resonant frequency. Measure or calculate the following values: $I_R =$ _____ mA;
$I_L =$ _____ mA; $I_C =$ _____ mA.

ANALYSIS

1. Calculate the percentage difference between measured and calculated values of resonant frequency of the original circuit:
Difference = _____ %.

2. Calculate the impedance of the original circuit at resonance by using the following formula:

$$Z_r = \frac{V_T}{I_T} = \underline{} \ \Omega$$

3. Calculate the quality factor of the original circuit:

$$Q = \frac{R}{X_L} = \underline{}$$

4. Calculate the bandwidth of the circuit:

$$BW = \frac{f_r}{Q} = \underline{} \ Hz$$

5. How do the values of I_R, I_L, and I_C vary as the frequency is moved to above and below resonance (see procedure steps 9–11)?
I_R: _____; I_L: _____; I_C: _____.

6. Complete the following statement: In a parallel resonant circuit at resonance, the impedance is _____, I_R is _____, the phase angle is _____, X_L is _____, and R is _____.

Frequency-sensitive AC Circuits

Unit 5 Examination

Frequency-Sensitive AC Circuits

Instructions: For each of the following, circle the answer that most correctly completes the statement.

1. If an ac source is connected in series with a resistor and a capacitor and a voltage output is taken across the capacitor, the circuit would exhibit the property of a

    a. Low-pass filter   b. High-pass filter
    c. Band-pass filter   d. Parallel resonant circuit

2. When frequency is increased in a low-pass filter circuit, the current flow will

    a. Increase because of greater X_L
    b. Increase because of less X_L
    c. Decrease because of greater X_L
    d. Decrease because of less X_L

3. At resonance, a series *RCL* circuit characteristically develops

    a. Maximum voltage across the input terminals
    b. Minimum current through the circuit
    c. Maximum reactance of the coil and capacitor
    d. Minimum impedance between the input terminals

4. Capacitive reactance is said to be frequency sensitive. This statement means that capacitive reactance:

    a. Is independent of frequency
    b. Decreases as frequency increases
    c. Increases with a rise of frequency
    d. Increases with a rise in capacitance

5. In a series resonant circuit, the impedance at resonance is equal to the resistance of the circuit.

    a. True   b. False

6. It is possible to pass a specific frequency through a resonant circuit by the proper selection of:

    a. *C* and *R*   b. *C* and *L*
    c. *R* and *L*   d. *R* and *Z*

7. At resonance, a parallel resonant circuit characteristically develops

    a. Maximum voltage across the input terminals
    b. Minimum current through the circuit
    c. Maximum reactance of the coil and capacitor
    d. Minimum impedance between the input terminals

8. When frequency is increased, voltage across a capacitor in a series *RC* circuit will

    a. Decrease   b. Increase
    c. Remain constant   d. Be infinite

9. In a series resonant circuit, when X_C and X_L are equal

    a. Line voltage leads line current
    b. Line current leads line voltage
    c. Total impedance is minimum
    d. Total impedance is maximum

10. If an ac source is connected in series with a resistor and a capacitor and a voltage output is taken across the resistor, the circuit exhibits the property of a

    a. Low-pass filter   b. High-pass filter
    c. Band-pass filter   d. Parallel resonant circuit

11. A circuit has a decibel gain of 12 dB. What is the power ratio associated with this gain (dB = 10 log P_1/P_2)?

    a. 1.58   b. 21.2
    c. 15.8   d. 2.12

12. Four kilohertz is equal to

    a. 400 cps   b. 4000 cps
    c. 40,000 cps   d. 4,000,000 cps

13. A 16 H inductor is connected in series with a 1 µF capacitor. What is the resonant frequency?

    a. 400 Hz   b. 126.5 Hz
    c. 12.65 Hz d. 40 Hz

14. A series resonant circuit has an f_r of 10 kHz and a Q of 10. The frequency range is

    a. 9.5 kHz to 10.5 kHz   b. 9 kHz to 11 kHz
    c. 1 kHz to 10 kHz   d. 10 kHz to 11 kHz

15. A series RCL circuit is at resonance when

    a. $X_L = 15\ \Omega$, $X_C = 5\ \Omega$, and $R = 15\ \Omega$
    b. $X_L = 10\ \Omega$, $X_C = 10\ \Omega$, and $R = 100\ \Omega$
    c. $X_L = 100\ \Omega$, $X_C = 5\ \Omega$, and $R = 5\ \Omega$
    d. $X_L = 20\ \Omega$, $X_C = 5\ \Omega$, and $R = 20\ \Omega$

APPENDIX A

Electronic Symbols

Fixed resistor	—⋀⋁⋀—	Ground	⏚ or ⏚	Fuse	—⌒⌒—
Tapped resistor	—⋀⋁⋀—	Contacts (normally closed)	—⫲—	Circuit breaker (single pole)	—⌢—
Variable resistor (potentiometer)	—⋀⋁↑—	Contacts (normally open)	—⫯—	Circuit breaker (three pole)	
Thermistor	—⋀⋁⋀— t°	Switch (single-pole, single-throw)	—o⟋o—	Coil (air core)	—⌒⌒⌒—
Fixed capacitor	—⊢⊣—	Switch (single-pole, double-throw)	—o⟋o—	Coil (iron core)	≡⌒⌒⌒≡
Variable capacitor	—⊢⊣↗—	Switch (double-pole, single-throw)		Coil (tapped)	—⌒•⌒—
Polarized capacitor (electrolytic)	—+⊢⊣−—	Switch (double-pole, double-throw)		Coil (adjustable)	—⌒↑⌒— or —⌒⌒↗—

Electronic Symbols 155

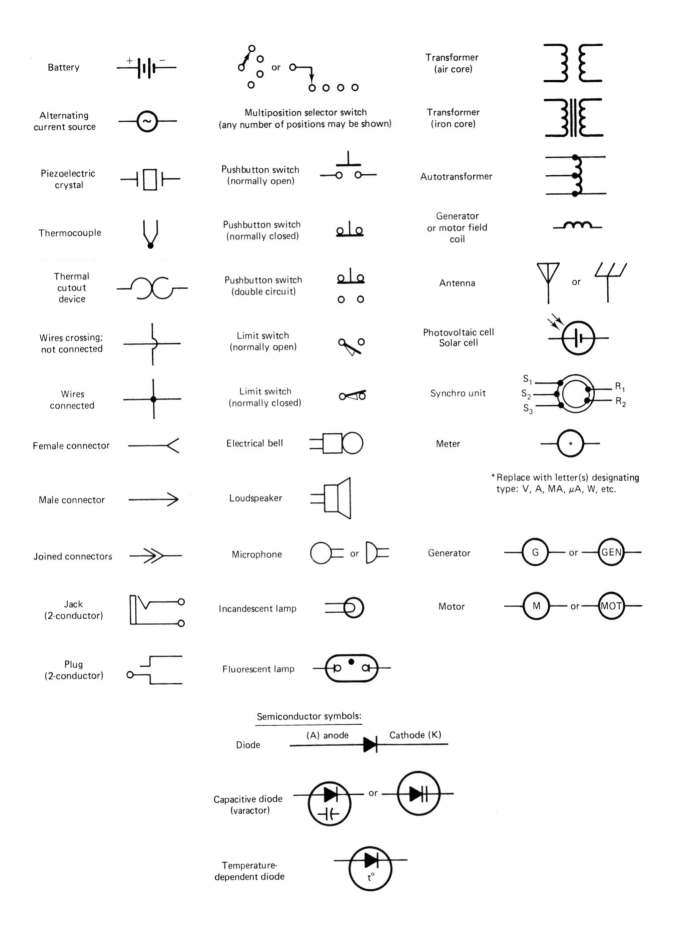

156 Appendix A

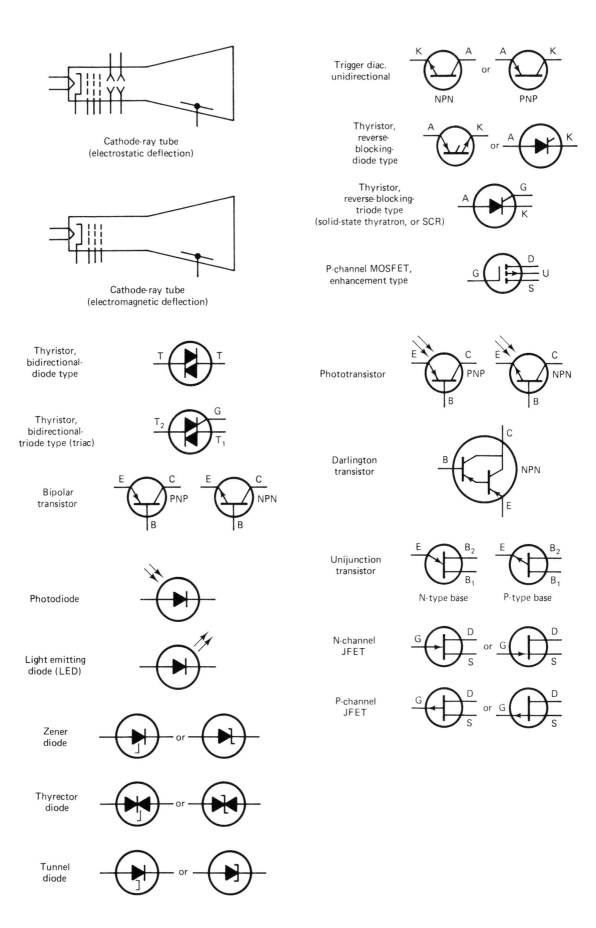

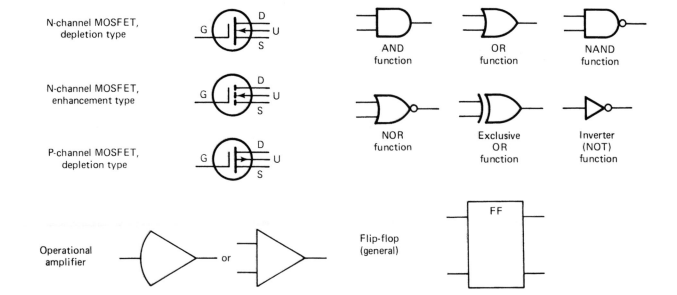

158 Appendix A

APPENDIX B

Trigonometry for AC Electronics

Trigonometry is a valuable mathematical tool for anyone who studies ac electronics. Trigonometry deals with angles and triangles, particularly the right triangle, which has one angle of 90°. An example of an electronics right triangle is shown in Fig. B-1. This example illustrates how resistance, reactance, and impedance are related in ac circuits. We know that resistance (R) and reactance (X) are 90° apart, so their angle of intersection forms a right angle. We can use the law of right triangles, known as the *Pythagorean theorem*, to solve for the value of any side. This theorem states that in any right triangle, the square of the hypotenuse is equal to the sum of the squares of the other two sides. We can express the Pythagorean theorem mathematically as follows:

$$Z^2 = R^2 + X^2$$

or

$$Z^1 = \sqrt{R^2 + X^2}$$

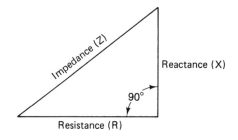
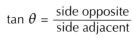
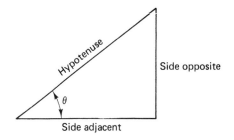

Fig. B-1. Right triangle shows trigonometric relations between resistance, reactance, and impedance in ac circuits.

Fig. B-2. Trigonometric relationships of the sides of a right triangle to angle θ.

By using trigonometric relations, we can solve problems dealing with phase angles, power factor, and reactive power in ac circuits. The three most used trigonometric functions are the *sine*, the *cosine*, and the *tangent*. Fig. B-2 illustrates how these functions are expressed mathematically. Their values can be found with a scientific calculator.

This process can be reversed to find the size of an angle when the ratios of the sides are known. The term *inverse function* is used to indicate this process. For example, the notation inv sin x=θ means that θ is the angle the sine of which is x. To solve inv sin 0.9455 = θ, look through the column of sine functions in Table B-1 and find that angle θ is 71°. These functions are easily solved with a calculator.

Trigonometric ratios hold true for angles of any size; however, angles in the first quadrant of a standard graph (0° to 90°) are used as a reference. To solve for angles greater than 90° (second-, third-, and fourth-quadrant angles), they must first be expressed as equivalent first-quadrant angles (Fig. B-3). All first-quadrant angles have positive functions, whereas angles in the second, third, and fourth quadrants have two negative functions and one positive function.

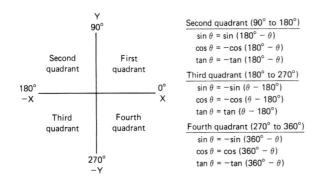

Fig. B-3.
Standard graph shows how trigonometric functions for angles greater than 90° (second-, third-, and fourth-quadrant angles) are derived by means of conversion to first-quadrant angles.

APPENDIX C

Electronic Equipment and Parts Sales

The following companies sell electronic equipment and parts. You can write to these companies to obtain catalogs and price lists for purchasing the equipment you need.

All Electronics Corp.
P.O. Box 567
Van Nuys, CA 91408-0567

Allied Electronics
7410 Pebble Dr.
Fort Worth, TX 76118

Brodhead-Garrett
100 Paragon Pkwy.
Mansfield, OH 44903
888-222-1332

Cal West Supply Inc.
31320 Via Colinas, Suite 105
Westlake Village, CA 91362
800-892-8000

Circuit Specialist Co.
P.O. Box 3047
Scottsdale, AZ 85257

Digi-Key Corp.
701 Brooks Ave.
Thief River Falls, MN 56701-0677

Edlie Electronics
2700 Hempstead Twp.
Levittown, NY 11756-1443
800-645-4722

ETCO Electronics
North Country Shopping Ctr.
Plattsburgh, NY 12901

F.W. Bell
6120G Hanging Moss Rd.
Orlando, FL 32807

Hewlett-Packard
1501G Page Mill Rd.
Palo Alto, CA 94305

Hickok Teaching Systems
2 Wheeling Ave.
Woburn, MA 01801
617-935-5850

Hughes-Peters
4865 Duck Creek Rd.
P.O. Box 27119
Cincinnati, OH 45227
800-543-4483

Jameco Electronics
1355 Shoreway Rd.
Belmont, CA 94002-4100

Kelvin Electronics, Inc.
1900 New Hwy.
P.O. Box 8
Farmingdale, NY 11735
800-645-9212

Lab Volt
Buck Engineering Co.
Farmingdale, NJ 07727

MCM Electronics
650 Congress Park Dr.
Centerville, OH 45459

Merlin P. Jones & Assoc.
P.O. Box 12685
Lake Park, FL 33403-0685
305-848-8236

Mouser Electronics
2401 Hwy. 287 N
Mansfield, TX 76063

Newark Electronics
3600 Chamberlain Ln.
Louisville, KY 40222-4397

Omnitron Electronics
770 Amsterdam Ave.
New York, NY 10025
800-223-0826

Priority One Electronics
21622 Plumer St.
Chatsworth, CA 91311
800-423-5922

RNJ Electronics
805 Albany Ave.
Lindenhurst, NY 11757
800-645-5833

Satco
924 S. 19th Ave.
Minneapolis, MN 55404
800-328-4644

Tektronix, Inc.
P.O. Box 1700
Beaverton, OR 97075

APPENDIX D

Soldering Techniques

Soldering is an important skill for electrical technicians. Good soldering is important for the proper operation of equipment.

Solder is an alloy of tin and lead. The solder used most is 60/40 solder. This means it is made from 60% tin and 40% lead. Solder melts at a temperature of about 400°F (205°C).

For solder to adhere to a joint, the parts must be hot enough to melt the solder. The parts must be kept clean to allow the solder to flow evenly. Rosin flux is contained inside the solder. It is called *rosin-core solder*.

A good mechanical joint must be made in soldering. Heat is applied until the materials are hot. When the materials are hot, solder is applied to the joint. The heat of the metal parts (not the soldering tool) is used to melt the solder. Only a small amount of heat should be used. Solder should be used sparingly. The joint should appear smooth and shiny. If it does not, it could be a cold solder joint, which is defective. Be careful not to move the parts when the joint is cooling. This could cause a cold joint.

When parts that can be damaged by heat are soldered, be very careful not to overheat them. Semiconductor components, such as diodes and transistors, are very heat sensitive. One way to prevent heat damage is to use a *heat sink*, such as a pair of pliers. A heat sink is clamped to a wire between the joint and the device being soldered. A heat sink absorbs heat and protects delicate devices. Printed circuit boards are also sensitive to heat. Care should be taken not to damage printed circuit boards when soldering parts onto them. Several types of soldering irons and soldering guns are available. Small, low-wattage irons should be used with printed circuit boards and semiconductor devices.

The following are rules for good soldering:

1. Be sure that the tip of the soldering iron is clean and tinned.

2. Be sure that all the parts to be soldered are heated. Place the tip of the soldering iron so that the wires and the soldering terminal are heated evenly.

3. Do not overheat the parts.

4. Do not melt the solder onto the joint. Let the solder flow onto the joint.

5. Use the right kind and size of solder and soldering tools.

6. Use the right amount of solder to do the job but not enough to leave a blob.

7. Do not move the parts before the solder joint cools.

APPENDIX E

Troubleshooting

Troubleshooting is a method of finding out why something doesn't work properly. If you follow logical steps, you will be able to locate most difficulties that occur in electronic equipment. Sometimes the trouble is so complex that it requires many hours of concentration and work. Other problems are easy to solve and require only a brief time.

Helpful resources for troubleshooting include the following:

1. Using a common sense approach
2. Knowing how electronic systems work
3. Knowing how to use test equipment
4. Knowing how to use schematics effectively
5. Being able to find the trouble through a logical sequence

To begin any kind of troubleshooting, determine possible courses of action. Without a system, the procedure of troubleshooting becomes a guessing game. You should be aware that no one system of troubleshooting is perfect. In the process of troubleshooting, keep in mind that most problems are usually component failures. If you know what each component is supposed to do, you will be aware of the troubles they can cause.

During troubleshooting, it is important that you use proper tests. Much of your time is used in locating the trouble. You must have a suitable approach to save time. As you become more familiar with troubleshooting, it becomes less time consuming.

As you continue your troubleshooting effort, you must constantly be aware of circuit or system operations you have already tested. Make a list or remember the probable troubles that have been tested.

Troubleshooting is a method of finding out why something does not work properly. If you follow logical steps, you will be able to locate most difficulties that occur in electronic equipment. Sometimes the trouble is so complex that it requires many hours of concentration and work. Other problems are easy to solve and require only a brief time. An important aspect of troubleshooting skill is initial inspection.

Initial inspection involves looking for the obvious. Several things should be done before actual circuit or system testing. In the initial inspection of any equipment, first open the equipment to look at it. You should look for the following:

1. Burned resistors. They are often obvious, may be charred, blistered, or bulged, and have discolored color bands or even holes.
2. Broken parts. These may come in the form of cracks, wires pulled out of parts, or destroyed parts.
3. Broken wires and poor connections.
4. Smoke or heat damage. Parts may smoke when equipment is turned on; this identifies defective parts but not the cause of the defect.
5. Oil leaks and water leaks.
6. Loose, damaged, or worn parts. These are found with visual and tactile examination.
7. Noisy parts. Uncommon noises indicate defective parts.

When initial inspection is performed properly, many troubles can be located without having to go through unnecessary steps. Initial inspection involves the senses of sight, touch, smell, and hearing. It is important to organize your thoughts to solve the problem. If you suspect a part of being the source of the problem, take a closer look. If you suspect a specific part, turn the equipment off and smell it, touch it, examine it closely. For example, a transformer that is good does not have an odor, but a burned transformer does. Initial inspection can help to locate the trouble in any defective electronic equipment or circuit.

The final solution of the problem involves application of your knowledge of electronic circuit operation and understanding the proper use of test equipment. Remember that troubleshooting is a *systematic* procedure.

APPENDIX F

Use of a Calculator

Electronic calculators greatly simplify problem solving for electronics technology. Low-cost calculators have the mathematical functions necessary to solve most problems encountered by electronics technology students. Most calculators have a simplified instruction manual that explains how to accomplish each mathematical function. The four basic functions (+, −, ×, ÷) are used for many simple electronics problems. Other functions that are particularly helpful in electronics applications are as follows.

Function	Key	Operation
Exponent	EE or EXP	Raises a number to a power of 10
Inverse	1/x	Divides 1 by a number
Square	x^2	Multiplies a number by itself (e.g., $5^2 = 5 \times 5$)
Square root	$\sqrt{x}$	Extracts the square root of a number
Trigonometric functions (sine, cosine, tangent)	sin cos tan	Determines the trig value of an angle (e.g., sin 30° = 0.5)
Inverse trig functions	inv sin inv cos inv tan	Determines the angle for a trig value (e.g., inverse sin 0.5 = 30°)
Logarithms	log	Finds the logarithm of a number (i.e., log 150 = 2.176)
Inverse logarithms	inv log	Finds the number associated with a logarithm (i.e., antilog 2.176 = 150)
Storage	sto	Stores a number in memory
Recall	rcl	Recalls a number from memory

Index

Active power, 71
Activities, experimental, 34–35
Admittance, 52
   triangle, 66
Air-core transformer, 105
Alternating current (ac), 1–3
   generating, 6–7
   single-phase, 7–8
   three-phase, 8–9
   voltage, 3–4
Alternator, 7–8
   three-phase, 8–9
Amplification, 116
Amplifier circuit, 3
Angle of lead or lag, 52
Antilogarithms, 126–127
Apparent power, 52, 66
Attenuation, 16, 18, 116, 127
Automobile alternator, 9
AUTO mode, 24, 31
Autotransformer, 106
Average voltage, 2, 4
Axis, 16

Band-pass filter, 116, 117, 118, 133
   experiment, 145–146
Bandwidth (BW), 116, 120, 121, 122
Basics, ac
   examination, 12–14
   self-examination, 10–11
Beam finder, 27
Bipolar transistors, 24, 25

Capacitance (C), 1, 51–54
   examination, 98–101
   experiment, 81–83
   self-examination, 73–74
   shunting, 25

Capacitive
   circuit, 58–60
   reactance, 52, 59, 117
      experiment, 81–83
Capacitor, 52
Cathode ray tube (CRT), 16, 17, 19–20, 26, 44
   deflection, 20–23
Center tap, 104
CHOP mode, 29
Circuits
   capacitive, 58–60
   differentiator, 135
   filter, 117–118
   inductive, 55–58
   integrator, 136
   parallel, 63–66
   resistive, 54–55
   resonant, 118–122
   series, 61–63
   See also Frequency-sensitive circuits
Coaxial cable, 25
Coefficient of coupling, 56
Compensating probe, 25–26
Competencies, lab activity, 36–39
Conductance (G), 52
Coordinate systems, 61
Counterelectromotive force (CEMF), 55
Current
   transformer, 106–107
   triangle, 64–66
Cutoff, 120
Cycle, 2, 6, 8
Cycles per second (cps), 3, 6, 8, 40

Decibel (dB), 116, 124
   applications, 125–127
   with filter circuits, 127–133

   self-examination, 137
   table, 130–132
Deflection, 16, 20–23
Delta connection, 2, 8–9
Dielectric, 52
   constant, 52
Differentiator circuit, 135
Direct current (dc), 3, 4, 17
   in oscilloscope, 24–25

Effective value, 2, 3, 4, 40
Efficiency, 107
Electrolytic capacitor, 53
Electromagnetic induction, 5–6
Electron
   beam, 16
   gun, 16, 19–20
Electrostatic field, 53
Examinations
   basics of ac, 12–14
   frequency-sensitive circuits, 152–154
   inductance and capacitance, 98–101
   measuring ac, 48–49
   transformers, 112–114
Experiments, 34–35
   band-pass filter, 145–146
   capacitance and capacitive reactance, 81–83
   high-pass filter, 142–144
   inductance and inductance reactance, 78–80
   low-pass filter, 139–141
   measuring ac voltage, 40–43
   oscilloscope measurement, 44–47
   parallel circuits
      RC, 95–97
      resonant, 150–151
      RL, 92–94

series circuits
    RC, 87–88
    resonant, 147–149
    RL, 84–86
    RLC, 89–91
transformer analysis, 110–111
EXT trigger, 24, 32

Faraday, Michael, 5
Farad (F), 53, 59
Filter circuits, 116, 117–118
    self-examination, 136
Focus, 17, 20, 26
Formulae
    bandwidth, 120, 121, 122
    capacitance, 58
    capacitive reactance, 59, 117
    capacitors in series and parallel, 60
    current, 54–55
    efficiency, 107
    frequency, 8
    frequency response, 129
    high-frequency cutoff, 120, 121, 122
    impedance, 53, 62
    inductive reactance, 53, 56
    instantaneous value to effective value, or vice versa, 40
    instantaneous voltage, 8
    line voltage, 9
    low-frequency cutoff, 120, 121, 122
    mutual inductance, 57
    parallel circuits, 64–66
    peak ac, 4
    peak-to-peak value, 40
    phase angle, 71
    power, 54–55, 66–72
    power amplification, 125–127
    power factor, 53
    primary and secondary voltage, 106
    quality factor, 120, 121, 122
    resistance, 54–55
    resonant frequency, 116, 119, 120, 121, 122
    series and parallel inductance, 57–58
    series circuits, 63
    susceptance, 54
    voltage, 54–55

Frequency, 2, 3, 6, 8, 40, 116
    response, 116, 132, 133
        curves, 117, 118, 127, 129
Frequency-sensitive circuits, 115–116
    decibels and, 124, 125–133
    differentiator circuit, 135
    examination, 152–154
    experiments, 139–151
    filter circuits, 117–118
    integrator circuit, 136
    resonant circuits, 118–122
    self-examination, 123, 136–138
    waveshaping control, 134–135

Gain, 17, 18, 124
Generated voltage source, 25
Generators
    electric, 7
    single-phase, 7–8
    three-phase, 8–9

Henry (H), 53
Hertz (Hz), 2, 3, 8, 40
High-cutoff frequency, 133
High-frequency cutoff, 120, 121, 122
High-pass filter, 116, 117, 132
    experiment, 142–144
Holdoff, 22
    time, 31
Home, power in, 3, 5, 7
Horizontal
    axis, 16
    gain, 18
    input, 18
    magnification, 30
    operating mode, 29–30
    position, 17, 29
    section, 19
    sweep, 18, 21, 23, 29
    time-base control, 30

Impedance (Z), 53
    triangle, 62
Inductance (L), 1, 51–54
    examination, 98–101
    experiment, 78–80
    self-examination, 72–73

Inductive
    circuit, 53, 55–58
    reactance, 53, 56
        experiment, 78–80
Inductors, 53, 56
Industrial and commercial power, 4, 9
In-phase, 2, 4
Instantaneous value, 2, 8, 40
Integrator circuit, 136
Intensity, 17, 26
INT trigger, 32
Inverse logarithms, 126
Iron-core transformer, 105
Isolation transformer, 104, 107

Junction field-effect transistor (JFET), 24, 25

Kilovolt-ampere (kVA), 66

Lab activity competencies list, 36–39
Lagging phase angle, 53
Leading phase angle, 53
Left-hand rule, 6
Line current, 9
LINE trigger, 24, 32
Line voltage, 9
Load, 1
Logarithms, 124–125
Low-cutoff frequency, 133
Low-frequency cutoff, 120, 121, 122
Low-pass filter, 116, 117
    experiment, 139–141

Magnetic field pole, 7
Mantissa, 124
Measured value, 3
Measurement, 15–16
    examination, 48–49
    experiments, 40–47
    multimeter, 17
    oscilloscope, 17–18
        operation, 18–32
    self-examination, 33–34
Mho, 53, 66

Mica capacitor, 53
Microfarad, 59
Multimeter, 3, 15, 17
Multiple-secondary transformers, 106
Mutual inductance (M), 53, 56–57, 104

Ohm's laws, 40
Oscilloscope, 15, 16, 18–19
    controls, 26–32
    CRT, 19–20
    deflection, 20–23
    experiment, 44–47
    measurement with, 17–18
    power supply, 24–25
    probes, 25–26
    triggering and synchronization, 23–24
Out of phase, 4–5

Parallel circuits, 63–66
    experiments, 92–97, 150–151
    resonant, 116, 122
    self-examination, 74–76
Parallel inductance, 57–58
Peak
    negative and positive, 3
    value, 2, 4, 40
Peak-to-peak, 2, 3, 28
Periodic wave, 134
Period (time), 2
Phase, 4–5
    angle, 2, 71
    current, 9
Phasor diagrams, 60–61
Picofarad (pF), 59
Polar coordinate system, 61
Poles, 8
Power
    factor, 53, 66, 71–72
    formulae, 54–55, 66–72
    generation, 5, 7
    ratings, 5
    ratio, 127
    supply, 19, 24–25
    transformer, 105
    triangle, 66, 70, 71
Powered metal-core transformer, 105
Primary winding, 103, 104, 105
Prime mover, 7

Probes, 16, 19, 25–26, 28
Pulse repetition rate (PRR), 135
Pulse waveforms, 134

Quality factor (Q), 116, 120, 121, 122

Radian, 54
Radio and television, 3, 105, 120
Ramp, 22, 29
Rating, voltage, 4
Reactance (X), 54
Reactive
    circuit, 54
    power, 54, 71
Rectangular coordinate system, 61
Repetitive pulse, 134
Resistance, 1, 51–54
Resistive-capacitive (RC) circuit, 59, 117
    experiments, 87–88, 95–97
Resistive circuit, 54–55
Resistive-inductive (RL) circuit, 55, 56, 117
    experiments, 84–86, 92–94
Resonance, 119
Resonant circuit, 116, 118–122
    experiments, 147–151
    self-examination, 137
Resonant frequency, 116, 119, 120, 121, 122
Resultant vector, 61
Retrace, 16, 22, 29
Revolutions per second (rev/s), 3, 8
Root-mean-square (RMS) voltage, 2, 4, 40
Rotor, 7, 8

Sawtooth wave, 16, 22, 23
Scope. *See* Oscilloscope
Secondary winding, 103, 104, 105
Selectivity, 116, 120
Self-examinations
    basics of ac, 10–11
    filter circuits, resonant circuits, decibel problems, 136–138
    frequency-sensitive circuits, 123
    measurement, 33–34
    resistance, inductance, capacitance, 72–77
    transformers, 107–109

Series circuits, 61–63
    experiments, 84–91, 147–149
    RC and RL, 117
    resonant, 116, 119–121
    self-examination, 74–76
Series inductance, 57
Shunting capacitance, 25
Siemen (S), 53, 54, 66
Signal-seeking mode, 31
Sine wave, 2, 3, 5, 7–8
Single-phase voltage, 2, 4–5, 7–8
Slip ring, 6, 7
Split ring, 6, 7
Stator, 7
Step-down transformer, 103, 104, 105–106
Step-up transformer, 103, 104, 105, 106
Susceptance (B), 54
Sweep, 16, 18, 21, 23, 29
Synchronization (sync), 16, 23–24, 30–31
    select, 18

Tank circuit, 116, 122
Television and radio, 3, 105, 120
Theta, 2, 61
Three-phase voltage, 2, 4–5, 7, 8–9, 107
Time
    axis, 17
    base, 16, 30, 31
Tools and equipment, 35
Trace, 16, 17, 29, 44
    rotation, 26–27
Transformer, 103–104
    applications, 107
    efficiency, 107
    examination, 112–114
    experiment, 110–111
    operation, 105
    self-examination, 107–109
    types, 105–107
Transient pulse, 134
Triangles
    circuit, 62, 63, 64, 65, 66, 67, 68, 69
    power, 66, 70, 71
Trigger
    coupling, 32
    level control, 23–24
    mode, 24, 31
    section, 19

signal sources, 31–32
Triggering, 16, 19, 23–24
   control, 30–31
Troubleshooting and testing competencies, 36–39
True power, 53, 54, 66, 71
Turns ratio, 104, 106

Unity
   coupling, 56
   power factor, 66

Variable
   holdoff control, 31
   time/centimeter, 18

time/division control, 30
volts/centimeter, 18
volts/division control, 28
Vector, 54
   diagrams, 60–61
Vertical
   attenuation, 18
   axis, 16
   deflection controls, 27
   gain, 17
   input, 18
   input coupling, 27–28
   operating mode, 28–29
   position, 17, 27
   section, 19
VERT MODE, 32
Voltage

axis, 17
   magnification, 121
Volt-amperes-reactive (VAR), 54, 71
Volt-ampere (VA), 52, 54, 66
Volt-ohm-milliammeter (VOM), 3, 15
Volts/division, 28

Wattmeter, 66
Watts, 54
Waveform, 3, 134
Wavelength, 3
Waveshaping control, 134–135
Working voltage, 54
Wye connection, 3, 8–9